THE NATURE OF ELECTROLYTE SOLUTIONS

£6.50

'D LOAN

43 654456

THE POLYTECHNIC OF WALES LIBRARY

Llantwit Road, Treforest, Pontypridd.

Titles in the *Dimensions of Science* series
Energy and Cells C. Gayford
Genes and Chromosomes J. R. Lloyd
Physics and Astronomy Donald McGillivray
Practical Ecology D. Slingsby and C. Cook
Ionic Organic Mechanisms C. Went
The Nature of Electrolyte Solutions Margaret Robson Wright

DIMENSIONS OF SCIENCE
Series Editor: Professor Jeff Thompson

THE NATURE OF ELECTROLYTE SOLUTIONS

Margaret Robson Wright

B.Sc., D. Phil.
formerly Lecturer in Chemistry,
now Honorary Lecturer in Chemistry,
The University, Dundee

MACMILLAN
EDUCATION

1021106 3

541.342
WRI

© Margaret Robson Wright 1988

All rights reserved. No reproduction, copy or transmission of this publication may be made without written permission.

No paragraph of this publication may be reproduced, copied or transmitted save with written permission or in accordance with the provisions of the Copyright Act 1956 (as amended), or under the terms of any licence permitting limited copying issued by the Copyright Licensing Agency, 33–4 Alfred Place, London WC1E 7DP.

Any person who does any unauthorised act in relation to this publication may be liable to criminal prosecution and civil claims for damages.

First published 1988

Published by
MACMILLAN EDUCATION LTD
Houndmills, Basingstoke, Hampshire RG21 2XS
and London
Companies and representatives
throughout the world

Printed in Hong Kong

British Library Cataloguing in Publication Data
Wright, Margaret Robson
The nature of electrolyte solutions.—
(Dimensions of science).
1. Electrolyte solutions
I. Title II. Series
541.3′72 QD565
ISBN 0–333–44427–2

10.2.89

In gratitude to my husband Patrick,
and also to
Andrew, Edward and Anne
and the cats

Series Standing Order

If you would like to receive future titles in this series as they are published, you can make use of our standing order facility. To place a standing order please contact your bookseller or, in case of difficulty, write to us at the address below with your name and address and the name of the series. Please state with which title you wish to begin your standing order. (If you live outside the United Kingdom we may not have the rights for your area, in which case we will forward your order to the publisher concerned.)

Customer Services Department, Macmillan Distribution Ltd
Houndmills, Basingstoke, Hampshire, RG21 2XS, England.

Contents

Series Editor's Preface

This book is one in a Series designed to illustrate and explore a range of ways in which scientific knowledge is generated, and techniques are developed and applied. The volumes in this Series will certainly satisfy the needs of students at 'A' level and in first-year higher-education courses, although there is no intention to bridge any apparent gap in the transfer from secondary to tertiary stages. Indeed, the notion that a scientific education is both continuous and continuing is implicit in the approach which the authors have taken.

Working from a base of 'common core' 'A'-level knowledge and principles, each book demonstrates how that knowledge and those principles can be extended in academic terms, and also how they are applied in a variety of contexts which give relevance to the study of the subject. The subject matter is developed both in depth (in intellectual terms) and in breadth (in relevance). A significant feature is the way in which each text makes explicit some aspect of the fundamental processes of science, or shows science, and scientists, 'in action'. In some cases this is made clear by highlighting the methods used by scientists in, for example, employing a systematic approach to the collection of information, or the setting up of an experiment. In other cases the treatment traces a series of related steps in the scientific process, such as investigation, hypothesising, evaluation and problem-solving. The fact that there are many dimensions to the creation of knowledge and to its application by scientists and technologists is the title and consistent theme of all the books in the Series.

The authors are all authorities in the fields in which they have written, and share a common interest in the enjoyment of their work in science. We feel sure that something of that satisfaction will be imparted to their readers in the continuing study of the subject.

Preface

Information about what goes on at the molecular level in an electrolyte solution tends not to be taught as an independent topic at the introductory level, but to appear incidentally in books which are essentially about electrochemistry. By electrochemistry I mean treatments which cover such topics as electrolysis, electrochemical cells, EMFs, weak acids and bases, conductivity, transport numbers and so forth. These treatments will normally cover basic theory and experimental techniques.

This book, on the other hand, is about the chemical species which are present in an electrolyte solution, and about the ideas which explain the properties and behaviour of electrolyte solutions at the molecular level. Of necessity, we are drawing on information derived from standard electrochemistry, and basic A-level knowledge of this subject is assumed in this book. However, as will become apparent in progressing through the book, thinking about the meaning of experiments on electrolyte solutions, and the interpretation of these experiments in terms of the nature and the behaviour of the solutions, requires drawing on information from branches of chemistry other than electrochemistry proper.

The nature of electrolyte solutions is not just a matter of interest only to electrochemists. A knowledge and understanding of what goes on in electrolyte solutions is vitally important to other chemists and scientists. For instance, many solution kineticists have to be as versatile and knowledgeable in electrolyte solution chemistry as they are in kinetics. Inorganic solution chemists would likewise limit themselves drastically if they did not have electrochemical ideas at their fingertips. And in the last decade or so it has become clear that electrochemical studies and concepts are becoming increasingly valuable and necessary in looking at many aspects of biological solutions.

This book, therefore, focuses attention on the current ideas about the nature and behaviour of the species present in aqueous electrolyte solutions. It also makes a very decided attempt to help and teach students how to think for themselves, and so the reasoning and the logic behind the topics discussed is emphasised. This is done by showing them how practising chemists go about thinking about behaviour at the molecular level, and how we set about interpreting experimental results and tying them up with theory or results from other branches of chemistry. The book also tries to teach students how to be critical of interpretative thinking by showing the pitfalls into which we could slip if we take too dogmatic an approach or interpretation. Once these points have been explicitly made, it should be easier for the student to see how thinking about what is happening in aqueous solutions can lead to further ideas and consequent advances in our knowledge.

When this has been done for one topic, here electrolyte solutions, I would hope that the student could transfer these thinking processes to other aspects of chemistry, or indeed of science, and thereby become a better all-round chemist or scientist.

It is unfair to expect students to work all this out for themselves, over and above the very necessary learning of facts. It is the teacher's duty to show and to explain how to think scientifically. The philosophy behind this book is that this is best done by detailed explanation and guidance, and by direct illustration on a limited scope of topics. This book, therefore, is not comprehensive. But if the ideas and thought processes outlined in the book are carried over to other topics studied elsewhere the student should find that 'he can do it for himself' and thereby can gain much more satisfaction and confidence than he would by just learning facts. It is understanding, being able to see for oneself, and confidence which help to stimulate and sustain interest.

Margaret Robson Wright
Department of Chemistry,
The University, Dundee

Acknowledgements

This book is the result of the accumulated experience of over twenty very stimulating years of teaching students at all levels. During these years I learned that being happy to help, being prepared to give extra explanation and to spend extra time on a topic could soon clear up problems and difficulties which many students thought they would never understand. Too often teachers forget that there were times when they themselves could not understand and when a similar explanation and preparedness to give time were welcome. If, through the written word, I can help students to understand and to feel confident in their ability to learn, and to teach them in a book in a manner which gives them the feeling of a direct contact with the teacher, then this book will not have been written in vain. To all the many students who have provided the stimulus and enjoyment of teaching I give my grateful thanks.

My thanks are due to Professor R. J. P. Williams, F.R.S., for reading the whole manuscript and sending encouraging comments.

In particular, I would like to express my very considerable gratitude to Professor Felix Franks who made a very careful and thorough reading of the whole manuscript. He made many extremely helpful and invaluable comments. These have been taken into account in the final version of the book, but the contents are my responsibility alone.

I would also like to thank Mary Waltham of Macmillan Education and Professor Jeff Thompson for encouraging me to write this book. Both helped enormously with their comments and advice throughout the writing.

My thanks are given to Mrs Irene Potter who did a first rate and very speedy rendering of the hand-written manuscript into a neat and well-presented typescript.

My husband, Patrick, has throughout my teaching career and throughout the thinking and writing of this book been a source of constant support and encouragement. His high intellectual calibre

and wide-ranging knowledge and understanding have provided ma
interesting and fruitful discussions. He has been an excellent sour
ing board for many of the ideas and manner of presentation in tl
book. He has read in detail the whole manuscript, and his clarity
insight and his considerable knowledge of the subject matter ha
been of invaluable help. My debt to him is enormous, and my grate
thanks are due to him.

1 Concepts and Ideas

This book is about the nature and behaviour of electrolyte solutions. This will include studying the types of particles which are present in such solutions, and will require reviewing the experimental evidence from which we infer the structure and nature of both the solute and solvent species. We will also be looking at the ideas involved in the theories which describe the nature of electrolyte solutions as well as studying the development of those theories.

But first of all it is useful to gather together in one place most of the facts and ideas which are pertinent and relevant to the understanding of electrolyte solutions. It also helps understanding if an explanation of how and why these facts are relevant is given, and this chapter sets this out for you.

ELECTROLYTE SOLUTIONS — WHAT ARE THEY?

Electrolyte solutions are solutions which can conduct electricity. Colligative properties such as the lowering of the vapour pressure, depression of the freezing point, elevation of the boiling point and osmotic pressure, all depend on the number of individual particles which are present in a solution. They can thus give us information about the number of particles **actually** present in the solution. For some solutes it is found that the number of particles **actually** present in solution is greater than would be expected from the formula of the compound.

In the study of electrolyte solutions, two types of solute particles can be distinguished:

(a) Those where the number of particles present is an integral number of times the number of particles expected on the basis of the stoichiometric unit, such as

$NaCl(aq)$: 1 stoichiometric unit $\longrightarrow$ 2 particles
$CaCl_2(aq)$: 1 stoichiometric unit $\longrightarrow$ 3 particles

and ***this ratio does not alter with change in concentration*** (see appendix A).

(b) Those where the number of particles is greater than that corresponding to the stoichiometric unit, but is much less than the values found in category (a), such as ethanoic acid, $CH_3COOH(aq)$, or ammonia, $NH_3(aq)$. Here the ratio of the **actual number** of particles to the **stoichiometric number** of the stoichiometric units ***increases dramatically with decrease in concentration*** (see appendix A).

The electrical conductance of aqueous solutions has been studied. Some are **non-conducting**, some **weakly conducting** and some **highly conducting**. Conduction of a current through a solution implies the existence of charged particles, and so conducting solutions must contain charged particles — ions. The highly conducting solutions correspond to solutions appearing in category (a), while the weakly conducting solutions correspond to category (b).

Crystal structures found from X-ray diffraction show that in the solid some solids consist of discrete molecular units while others are giant lattices of ions held together by strong electrostatic interactions, and with no one cation specifically belonging to any particular anion, and vice versa.

The solutes whose structures in the solid is a giant ionic lattice are those which give strongly conducting solutions and whose colligative properties place them in category (a). Colours of ionic solutions are also indicative of individual charged particles being present, for instance, copper salts are always blue, dichromate salts are orange.

The **conclusion** to be drawn from these studies is that in solutions the solute can exist as

(a) **Molecular units**: non-conducting, *normal* colligative properties, X-ray structure showing discrete molecular units in the solid.

(b) **Molecular units plus ions**: weakly conducting, colligative properties showing **slightly** more than the expected numbers of particles present, X-ray structure showing discrete molecular units in the solid.

(c) **Ions**: highly conducting, colligative properties considerably greater than expected, X-ray structure showing a giant ionic lattice.

Solutes giving in solution:

(a) **Molecular units** are called **non-electrolytes**
(b) **Molecular units plus ions** are called **weak electrolytes**
(c) **Ions only** are called **strong electrolytes**

For the **weak electrolytes** the molecules are present in equilibrium with ions derived from the molecules:

$$\text{Molecules} \rightleftharpoons n \text{ ions}$$

$$K_{\text{dissoc}} = \left(\frac{[\text{ions}]^n}{[\text{molecules}]}\right)_{\text{equilibrium}}$$

Determination of values of the equilibrium constant has been an important experimental study, with conductance and EMF methods being the main tools.

For a **weak electrolyte** which is only partially dissociated into ions at moderate concentrations, there is a dramatic increase in the fraction of ions present as the concentration decreases. This results in a dramatic increase in the molar conductance as the concentration decreases.

A **strong electrolyte** consists of ions with no significant amounts of molecular species present. We would expect that the molar conductance would be independent of concentration. (Molar conductance and related electrical quantities are given in appendix A.)

With extensive study it was soon found that anomalies existed, for instance:

(a) For **strong** electrolytes the molar conductance is **not** independent of concentration.
(b) Electrolytes which are **strong** in **aqueous** solutions are shown to behave like **typical weak electrolytes** in **solvents** like **dioxan, acetone** or **methanol**.
(c) These anomalies are reflected in other studies using strong and weak electrolytes.

This forced chemists into focusing their attention on two main points:

(i) What exactly is an electrolyte solution like, and how does it behave at the **molecular** level? In particular, because ions are charged particles, do electrostatic interactions play a part in the **observable** behaviour of electrolyte solutions?
(ii) What exactly is the role of the solvent, and do we have to consider its **molecular** properties as well as its **bulk** properties?

I. IONS — SIMPLE CHARGED PARTICLES OR NOT?

Some simple basic properties of ions are regularly used in our discussion of electrolyte solutions and in theories describing the behaviour of electrolyte solutions. These ideas are often physically naive and we will find that we must modify them before we can ever hope to have a physically realistic description of electrolyte solutions.

These properties will be summarised below with indications of which are naive. A very brief indication of where modifications will be needed is given after the summary. Full discussion of these points is dealt with in later chapters.

Simple Properties of Ions

1. Ions have **integral** positive or negative charges.
2. Ions have finite definite sizes
   — **but** see discussion below of **solvation**.
3. Ions are often considered to be spherically symmetrical
   — **but** see discussion below of **shapes** of ions.
4. We normally consider the charge to be evenly distributed over the surface of the ion
   — **but** see discussion below of **charge-separated** ions.
5. We normally talk of ions as being unpolarisable
   — **but** see discussion below of **polarising power** and **polarisability**.
6. We normally think of each ion as moving as an independent entity
   — **but** see discussion below of **ion-pairing** and **micelle clustering**.

7. Ions can be ordered in terms of their ease of discharge at an electrode.
8. H^+ and OH^- in aqueous solutions show special properties.

Modifications needed to those Simple Ideas: a Summary

Sizes. Sizes of ions in solids are found by X-ray crystallography and are termed 'crystallographic radii'. These radii are often used in discussions of properties of electrolyte solutions. But they really should not be used for ions in solution. Many studies indicate that bare ions rarely exist in solution, and that their effective size is a combination of the crystallographic radii plus a contribution from solvation effects. Electrochemical experiments can yield solvation numbers, but the main evidence comes from other studies which will be discussed in chapter 4. Knowledge about solvation of ions is vital to our understanding of the behaviour of electrolyte solutions and has proved to be of crucial importance in determining the behaviour of biological systems. Because of the considerable current interest in this topic, a full chapter will be devoted to solvation.

Shapes. Because so much of the theoretical discussion of electrolyte solutions is based on an assumption that ions are spherically symmetrical, there is a tendency to forget that many ions are certainly not spherically symmetrical. A conscious effort should be made to think about the shape of an ion, as well as its charge.

Ions such as Ca^{2+}, Mg^{2+}, Cl^-, SO_4^{2-} and PO_4^{3-} are nearly spherically symmetrical, as are many of the complex ions found in inorganic chemistry:

$$Cu(NH_3)_6^{2+} \qquad Fe(CN)_6^{4-}$$
$$Ni(CH_2(COO)_2)_2^{2-} \qquad AlCl_6^{3-}$$

but many organic ions are not:

$$\overset{+}{N}H_3CH_2CH_2COOEt \qquad C_6H_5\text{—}CHBrCHBrCOO^-$$

while complex ions like those which are often found in biologically active solutions generally are non-spherical:

$$
\begin{array}{l}
\text{NH}\ \ \text{CH}_3 \\
\ \| \qquad | \\
\text{C}—\text{N}—\text{CH}_2\text{COO}^- \\
| \\
\text{NH} \\
| \\
\text{PO}_3^{2-}
\end{array}
\qquad
\begin{array}{l}
\overset{+}{\text{N}}\text{H}_3\text{CH(CH}_2)_2\text{S(CH}_2)_2\text{CH}\overset{+}{\text{N}}\text{H}_3 \\
\qquad\ \ | \qquad\qquad\qquad\quad | \\
\quad\ \text{COO}^- \qquad\qquad\ \ \text{COO}^-
\end{array}
$$

In fact, the **shape** of these ions often plays a very important role in the biological activity, and so biologists and biochemists must always remember that the physical chemist's theories on electrolytes may often have to be modified before they are relevant to the sorts of electrolytes normally encountered in biological systems.

Distribution of charge on an ion. Even distribution of charge over the surface of an ion is a valid assumption for simple spherical ions such as Ca^{2+}, SO_4^{2-} and $AlCl_6^{3-}$, but is most probably invalid for ions such as

$(CH_3)_3\overset{+}{N}CH_2CH_2CH_2CH_3$ $\qquad$ $CH_3CH_2CH_2COO^-$

$—SO_3^-$ $\qquad$ $SHCH_2CH\overset{+}{N}H_3$ (with COO^- on the CH)

Furthermore, many organic ions, often associated with biologically important substances, have a total charge which is a multiple of one, but where this charge is made up of individual charges at different sites in the molecule as in

$(CH_3)_3\overset{+}{N}CH_2CH_2\overset{+}{N}(CH_3)_3$	$SO^-{}_3CH_2CH_2CH_2SO_3^-$
$\overset{+}{N}H_3CH_2COO^-$	$\overset{+}{N}H_3(CH_2CONH)_nCH_2COO^-$
$SO^-{}_3CH_2CH_2COO^-$	proteins
polyphosphates	nucleotides
	polysaccharides

It is quite obvious that even distribution of charge over these ions simply will not occur, but a further very important factor must also be considered. Do these substances behave as ions with a given net overall charge and can they be treated as though they were equivalent to simple ions such as Ca^{2+}, SO_4^{2-}? Or do they behave as though each individual charge simulated a separate independent ion? Evidence suggests that in certain solvents and above certain concentrations, effects such as these are highly pertinent to the understanding of the behaviour of these electrolytes.

Again, it is vitally important that biologists and biochemists are aware that the electrolyte behaviour of a charge-separated ion is likely to be radically different from that of simple ions, and that ideas and theories developed for the simple situations will have to be modified in the biological context. A fuller treatment will appear in a later chapter.

Unpolarisable ions: It is clearly untrue to consider ions to be unpolarisable, but theoretical treatments generally discuss ions as though they were unpolarisable.

However, even the simple I^- ion is highly polarisable. This is a well accepted fact in other branches of chemistry, for instance, Fajans' Rules in inorganic chemistry deals explicitly with this effect in bonding, and in summary we can say

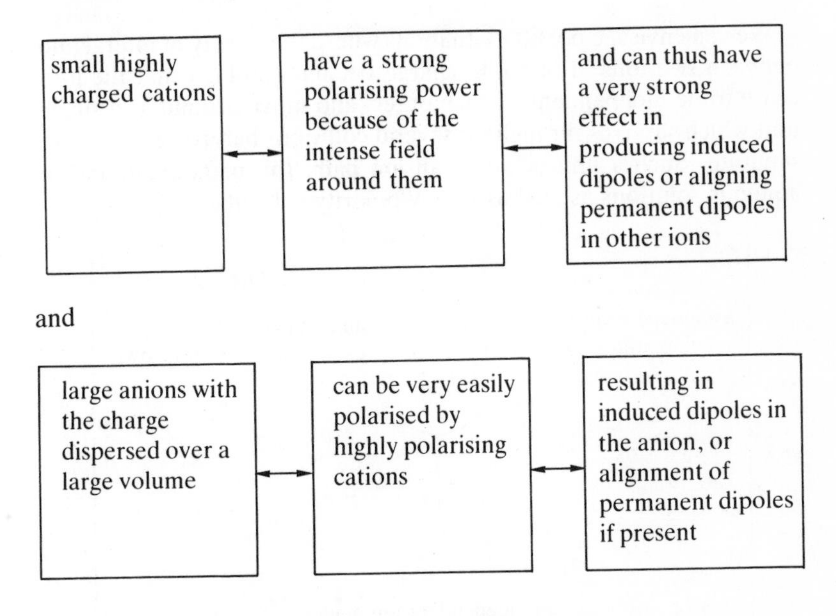

small highly charged cations
have a strong polarising power because of the intense field around them
and can thus have a very strong effect in producing induced dipoles or aligning permanent dipoles in other ions
and
large anions with the charge dispersed over a large volume
can be very easily polarised by highly polarising cations
resulting in induced dipoles in the anion, or alignment of permanent dipoles if present

These effects are present in the simple ions of standard inorganic substances such as Cs^+, Ca^{2+}, Cl^-, SO_4^{2-}, NO_3^-, CH_3COO^- and $(CH_3)_4\overset{+}{N}$. They are also going to be of considerable importance in electrolyte solutions where many of the ions are large and complex, for example protein ions, phospholipid ions, nucleic acids, ions of neurotransmitters and so on. (Dipoles, induced dipoles and alignment of dipoles are discussed later.)

Complete dissociation into ions: We shall see later that one of the main functions of the solvent is simply to reduce the forces of interaction between ions and thereby reduce the electrostatic potential energy of this interaction. Physically this corresponds to allowing the ions to exist as ions. This is a bulk effect.

Experiments and theoretical considerations which will be described later tell us that:

(a) Increasing the concentration of the solution increases the electrostatic interaction energy between the ions.
(b) Solvents of low polarity result in a greater electrostatic interaction energy between the ions compared with the electrostatic interaction energy between the ions when they are in solvents of high polarity.

We can envisage possible situations where the energy of interaction between two ions of opposite charge becomes so high that the ions cease to be independent of each other, and move around as a single unit which survives throughout several collisions before being able to separate. Such a unit is called an **ion pair**. Ion pairs are found in aqueous solutions as well as in low polarity solvents.

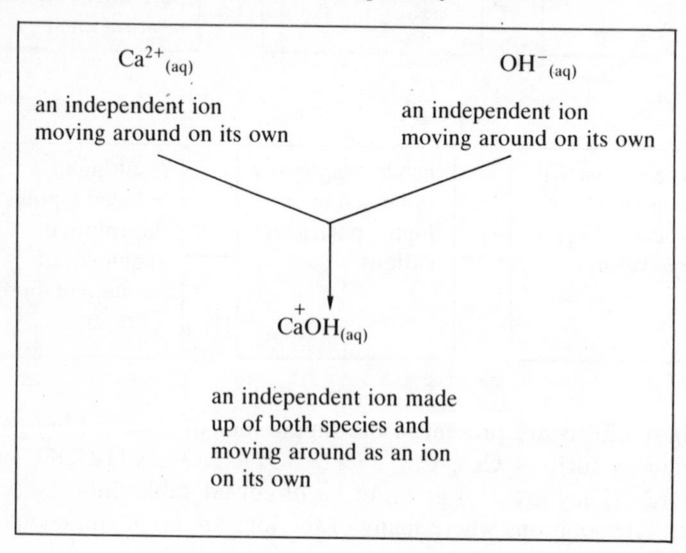
$Ca^{2+}_{(aq)}$
an independent ion
moving around on its own
$OH^{-}_{(aq)}$
an independent ion
moving around on its own
$\overset{+}{Ca}OH_{(aq)}$
an independent ion made
up of both species and
moving around as an ion
on its own

This is a very important topic in electrolyte studies and will be referred to often throughout the book.

Ions can also be formed into **clusters** called **micelles**, and this can become very important in colloidal solutions.

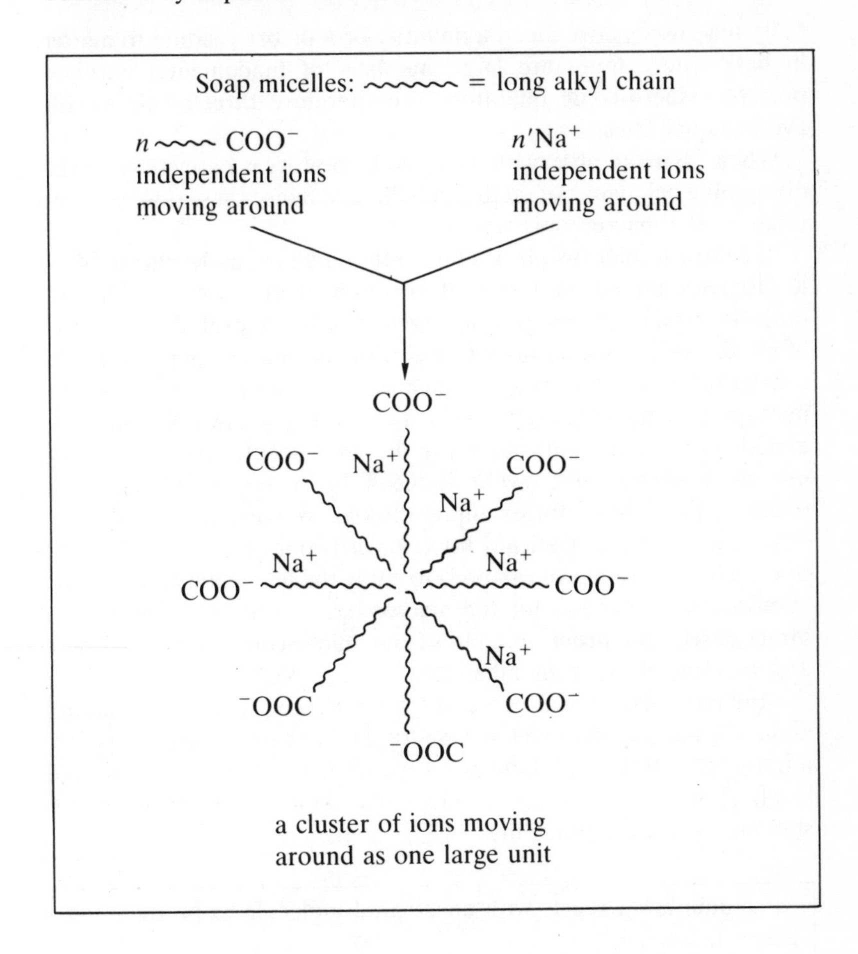

II. SOME NEW BUT NECESSARY TERMS

An understanding of electrolyte solutions implies an awareness of just what is present at the **microscopic** level. We also need to know how the particles we believe to exist at this level behave physically and chemically.

The first idea which we must get absolutely clear is

What do we mean by the terms macroscopic and microscopic?

By **macroscopic** we mean a quantity or a theory relating to **matter in bulk** where there are large numbers of fundamental particles present. **Macroscopic** quantities are therefore **directly observable average** quantities.

When we make observations in the laboratory we are studying the observable behaviour of matter in bulk, and hence these observations relate to the macroscopic reality.

In contrast, **microscopic** is a term relating to the **molecular level** — it discusses matter in terms of the individual atoms, molecules, radicals, ions, electrons, protons, neutrons. It is a level of description of matter which is generally **inferred** from our macroscopic level. It is a level which we cannot see, touch or manipulate directly, except, perhaps, in some very highly refined experiments. And even in these experiments we are still observing the microscopic level by making use of a macroscopic event induced by a single molecular or microscopic event — for example, radioactive counts.

We cannot ever realistically hope to verify our microscopic descriptions. The only thing we can do is to study the implications which our microscopic reality has for the macroscopic quantities. There is no **direct observable proof** possible of our microscopic interpretations, and we must always remember this.

Familiarity can breed contempt, and we get so used to talking about the microscopic level that we forget that what we are saying is a microscopic statement about what we observed at the macroscopic level. How often do we stop and think about this when we make statements like the following?

"a sodium atom reacts with an ethanol molecule to produce the ethoxide ion",

or

"ethanoic acid is only partially ionised into hydrated protons and ethanoate ions",

or

"methane reacts with chlorine molecules via a reaction mechanism which involves chlorine atoms, methyl radicals, chloromethyl radicals and hydrogen atoms and so forth".

One aim of this book is to make you fully aware of the inter-relation of the microscopic description with the observable macroscopic phenomenon, and to make you conscious of just how much inference has gone into the build up of the whole atomic, molecular and ionic descriptions which constitute chemistry. Such considerations lie at the heart of scientific methods.

Macroscopic and Microscopic Theories — a Distinction

A **macroscopic** theory is one dealing with the **directly observable world** of chemistry and is totally independent of any postulate of the existence of atoms, molecules, ions and such particles. Thermodynamics is the main macroscopic theory, and it develops equations describing observable quantities in terms of the bulk properties of matter.

A **microscopic** theory is one describing how matter behaves at the molecular level. It attempts to formulate equations between observable macroscopic quantities, and does so from a model which is based on inferred microscopic particles such as atoms, molecules, radicals or ions, and from what **we think** their behaviour is at the molecular level. Theories such as electrolyte theory, internal structure of crystals, reaction mechanisms and quantum theory are examples of microscopic theories.

An **interlinking theory** joins up these two extremes so that the final description which we have is one which is both **macroscopic and microscopic**. Statistical mechanics is the major theory in chemistry lying in this category — it is microscopic in so far as events are described at the molecular level; it is macroscopic in so far as the equations describing these events are often wholly or partly in terms of macroscopic quantities, and because thermodynamic theory is often intimately built into the description.

Quantities — Which Ones are Macroscopic or Interlinking?

Macroscopic quantities. Typical ones are temperature, pressure, volume, molar conductance, rate constant, concentration, relative permittivity, viscosity, and **all** thermodynamic quantities such as equilibrium constants, enthalpy, free energy, molar heat capacity and so on.

Microscopic quantities. Typical ones are ionic conductance, crystallographic radii, hydrodynamic radii, bond dissociation energy, quantum numbers, internuclear distance and so forth.

Interlinking quantities. The main one is the partition function of statistical mechanics which is a device for linking microscopic quantities with macroscopic quantities. Some basic interlinking quantities are used in computer simulations. But most interlinking quantities are beyond the scope of this book. They are mentioned so as to make you aware that such quantities do exist.

A Summary

(i) **Macroscopic** — observable, bulk.
(ii) **Microscopic** — not observable directly, molecular and inferential.
(iii) **Remember** — nearly every descriptive statement which we make about chemical behaviour and chemical reactions is very often almost a totally microscopic description.
(iv) **Remember** — when we talk about quantities and experimental measurements and results we are talking at a macroscopic level.

III. THE SOLVENT: STRUCTURELESS OR NOT?

The solvent is the **medium** in which the solute exists. It is often called a **dielectric**. A **dielectric** can be thought of in terms of an **insulator**, which is a substance which stops or tends to stop the flow of charge, in other words to stop a current passing through it.

If a substance which acts as an insulator is placed between two charges, it reduces

(a) the field strength,
(b) the force acting between the charges,
(c) the electrostatic potential energy of interaction between the charges,

and the factor by which it reduces these quantities is the relative permittivity ϵ_r. (These quantities are discussed in appendix B.) It is very important to realise that this definition of the relative permittivity is independent of any assumption as to what the dielectric is made of; in particular, it is independent of any assumption that the dielectric is composed of atoms and molecules, and so requires no discussion of the medium at a **microscopic** level. In effect, the relative permittivity ϵ_r is just a proportionality constant characteristic of the medium.

This is precisely what is meant when we talk about the medium being a **structureless** dielectric or **continuum**. In particular, when we are talking about the role of the solvent in electrolyte solutions, we are constantly referring to it as a continuous medium, or a structureless medium. Most of the theoretical discussions of electrolyte solutions formulate the theoretical equations in terms of factors which involve the bulk macroscopic quantity ϵ_r. Use of this bulk quantity in the equations implicitly means a description in terms of the solvent being a structureless dielectric with no microscopic structure which need concern us.

However, the use of a macroscopic quantity in the equations does not preclude us from discussing whether it is reasonable or not, or indeed sensible, to consider the solvent as having a purely bulk macroscopic role to play. From other chemical studies we know that the solvent is made up of molecules with a certain microscopic structure, so it is perfectly reasonable to expect that the microscopic structure may be of vital importance when the solvent plays its role as solvent in electrolyte solutions.

Indeed

it is precisely by addressing ourselves to this questions that vast progress can be made in our understanding of electrolyte solutions. We now realise that molecular details of ion–solvent interactions and consequent modifications to solute–solute and solvent–solvent interactions make a contribution to the behaviour of electrolyte solutions.

Studies of the microscopic structure of the solvent and its modification by the ions of the electrolyte have resulted in considerable refinements being forced on to our model of electrolyte solutions. Unfortunately, it is much easier to alter the model to incorporate new ideas and thoughts than it is to incorporate these ideas into the mathematical framework of the theory of electrolyte solutions and its derivation. We shall be looking at the implications of many of the topics introduced in this chapter when we look at theories of electrolytes (chapter 3) and solvation (chapter 4).

IV. THE MEDIUM — ITS STRUCTURE AND THE EFFECT OF IONS ON THIS STRUCTURE

We can ask ourselves the question:

> Does the fact that the medium (solvent) reduces the effect of one charge on another charge mean that the charges on the ions must have some effect on the medium?

To answer this question we must look at the details of the molecular structure of the medium, here the solvent. At this stage we will only briefly introduce some important ideas which will be developed in more detail in the chapter on solvation.

The solvent is made up of molecules which are in turn made up of nuclei and electrons. In covalently bonded molecules the bonds are formed by sharing of electrons between the two atoms involved in the bond.

> A bond is termed **non-polar** if the electron distribution in the bond is **symmetrical**, that is, the atoms have an **equal** share of the electrons of the bond.

Examples are

$$\mathrm{Cl-Cl;\ H-H;\ N\equiv N;\ C-C}$$

> A bond is termed **polar** if the electron distribution in the bond is **asymmetric**, that is the atoms have an **unequal** share of the electrons of the bond.

Examples are

$$\mathrm{C-Cl} \text{ written } \overset{\delta+}{\mathrm{C}}-\overset{\delta-}{\mathrm{Cl}} \text{ or } \overrightarrow{\mathrm{C-Cl}}$$

$$\mathrm{C=O} \text{ written } \overset{\delta+}{\mathrm{C}}=\overset{\delta-}{\mathrm{O}} \text{ or } \overrightarrow{\mathrm{C=O}}$$

$$\mathrm{H-F} \text{ written } \overset{\delta+}{\mathrm{H}}-\overset{\delta-}{\mathrm{F}} \text{ or } \overrightarrow{\mathrm{H-F}}$$

$$\mathrm{S=O} \text{ written } \overset{\delta+}{\mathrm{S}}=\overset{\delta-}{\mathrm{O}} \text{ or } \overrightarrow{\mathrm{S=O}}$$

When the bond is polar we give it a **bond dipole moment** which, on a qualitative basis, describes the degree to which the electron distribution is asymmetric. The direction in which the electron density is highest is shown by the head of an arrow

$$\overrightarrow{C - Cl}$$

and an indication of the relative displacement of positive and negative charges is given as

$$\overset{\delta+}{C} - \overset{\delta-}{Cl}$$

But does this tell us anything about the polarity of the molecule as a whole?

Answering this means looking at the overall effect of the bond dipoles, and this requires knowledge of the symmetry of the arrangement of the atoms in the molecule.
Several conclusions follow:

(a) If **all the bonds** in the molecule are **non-polar**, or **virtually non-polar** (as in C—H), then the molecule as a whole will be **non-polar**. For example

H_2, N_2, O_3, S_8, alkanes, benzene

(b) But if **some of the bonds** are **polar**, then the molecule as such will be
   (i) **non-polar** if the arrangement of the atoms is **symmetrical** and the vector bond dipole moments cancel out, as in CCl_4
   (ii) **polar** if the arrangement of the atoms is **not symmetrical** and the vector bond dipole moments do not cancel out, as in CH_3Cl, H_2O, CH_3OH.

Now can these ideas be used to help us think out what might happen when an ion is put into a solvent?

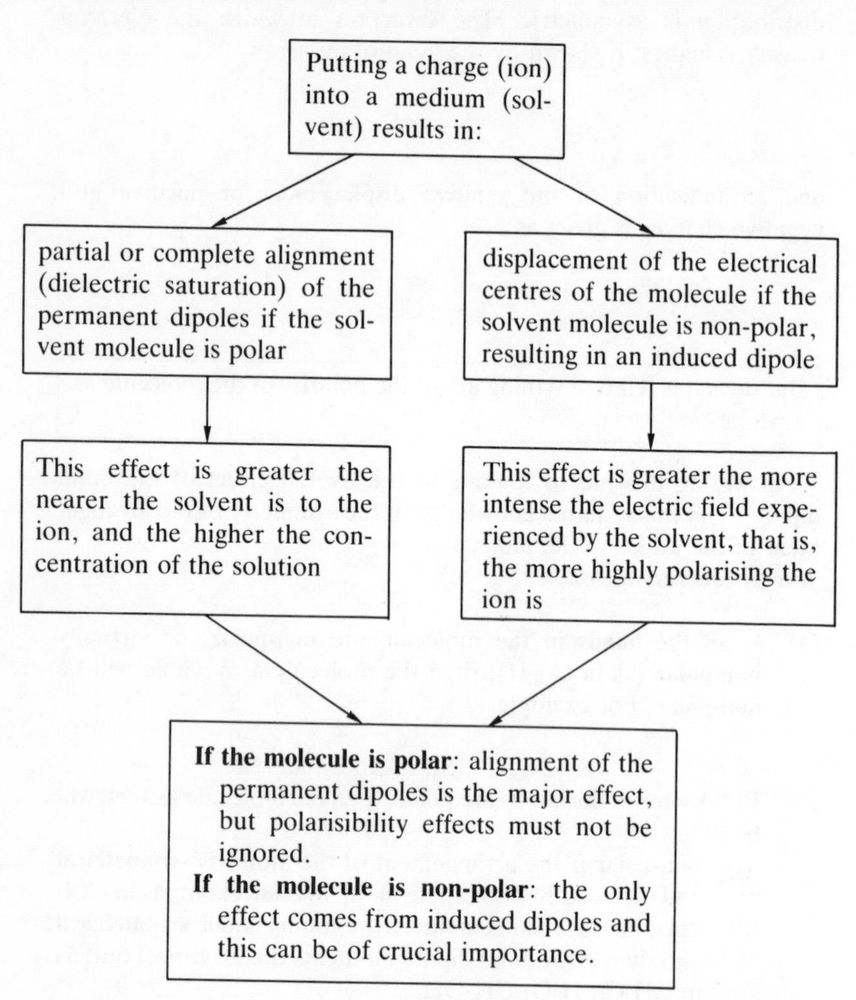

Both these effects are of great importance when we consider ion–solvent interactions in solution. In particular, their existence implies that we must acknowledge that ions **do modify** solvent structure. This, in turn, implies that there must be a **consequent modification** of solute–solute interactions in particular, and also of solvent–solvent interactions though these are of less importance.

We have seen that the relative permittivity of the solvent represents the effect of the solvent on the ion–ion interactions. This effect is a result of the ability of the ions to align the permanent dipoles or to induce dipoles in the solvent molecules. The more the dipoles are aligned by the ions the smaller is the value of the relative permittivity, ϵ_r, since this is measured experimentally by observing the effect of an external field on the solution. If the ions have totally aligned all the dipoles in the solvent molecules, there will be none left for the external field to align. The net result is that the only effect that the external field can have will be to induce further dipoles — a much smaller effect. Hence the measured relative permittivity will be low for such a situation.

Furthermore the possibility of alignment of the dipoles of the solvent molecules by the ions leads us to the following conclusion. For the solution, the simplest model to be envisaged is one where three possible situations can be thought of. And for each situation we must ask ourselves a question.

1. There is a region close to the ion where all the permanent dipoles are **completely aligned** — the region of dielectric saturation.
Question: Can it possibly be valid to use the macroscopic relative permittivity of the **pure** solvent to describe this region? Yet we do this in most of our theoretical treatments.

2. There is a region where **partial** alignment occurs, and the situation changes through the region from **complete** alignment to one where there is **non-alignment**.
Question: What value should we assign to the relative permittivity here, and should it vary throughout the region?

3. There is a region where almost **no alignment** occurs and where the solvent behaviour approximates to that of the pure solvent.
Question: Is it legitimate to use the bulk relative permittivity of the pure solvent here?
The answer is probably yes.

From this very basic and elementary discussion it should be abundantly clear that the solvent plays a crucial role in the behaviour of electrolyte solutions quite apart from its role as a dielectric reducing the forces of interaction between ions. Such considerations are of vital importance in physical chemistry and have really only been tackled rigorously in the last decade or so. Where complex electrolytes such as those encountered in biology are concerned, they are of crucial importance and may well dominate the behaviour of such solutions. Unfortunately, for these complex solutions, our understanding here is only in its infancy.

V. ELECTROSTRICTION

In the vicinity of each ion, a certain shrinkage of the solvent is likely to occur as a result of the attraction between the ionic charge and the polar molecules. This is called **electrostriction**, and leads to a local increase in the density of the solvent around each ion since more solvent molecules will be packed into the volume around the ion than would be present in that volume were the ion not present.

Electrostriction is important in solvation, but has not ever been properly incorporated in any detail into electrolyte theory.

VI. IDEAL AND NON-IDEAL SOLUTIONS — WHAT ARE THEY?

Ideality is a concept which can be used for a **pure** substance only if the interactions between the particles are negligible as in a **gas** at low pressures. Since **pure liquids** and **pure solids** are **condensed** phases, they must have significant forces of interaction between their fundamental particles. It is therefore meaningless to talk about ideality for either the pure liquid or pure solid. But ideality and non-ideality are important when talking about mixtures.

If we think of a two-component mixture this can alter in composition from a situation where the amount of component A is zero to one where the amount of component B is zero. In between these two limiting situations we can have mixtures where there are varying proportions of the two components. A mixture where the amount of one component **tends** to zero corresponds to the **ideal** mixture. Mixtures where there are **finite** amounts of both components present correspond to **non-ideal** mixtures.

Electrolyte solutions are mixtures where the components are the solute and the solvent. When the concentration of the solute **tends** to zero the solution is regarded as **ideal**. When there are **finite** concentrations of solute the solution is regarded as **non-ideal**.

In electrolyte solutions there will be interactions between

solute–solute particles
solute–solvent particles
solvent–solvent particles

In electrolyte solutions the solute is partly or wholly in the form of ions in solution. Because ions are charged particles there will be **electrostatic** interactions between the ions, and between the ions and the solvent over and above the solvent–solvent interactions.

A. Solute–Solute Interactions, that is, Ion–Ion Interactions

These basically are made up of four contributions.

(a) **Long-range-coulombic interactions** — that is, those acting over long distances between the ions. These are electrostatic interactions obeying the Coulomb inverse square law:

$$\text{force} \propto r^{-2} \text{ (see appendix B)}$$

(b) **Ion–induced dipole interactions** if the ions are polarisable, as indeed most ions are (see section C below for discussion of induced dipoles).
(c) **Short-range interactions** — that is, those acting over short distances. These can be coulombic or non-coulombic in nature.
(d) **Short-range repulsions**, which become significant at short distances between the ions. These are always present when two particles approach close to each other.

B. Solute–Solvent Interactions — that is, Ion–Solvent Interactions — Collectively known as Solvation

These interactions are made up of

(a) ion–dipole interactions
(b) ion–induced dipole interactions (see section C below for a description of these effects).

These interactions are **attractive** when the interaction is between the end of the dipole which has opposite charge to that of the ion with which the dipole or induced dipole is interacting.

They are **repulsive** when the ion is in close enough proximity to the end of the dipole which has a charge the same as the ion.

C. **Solvent–Solvent Interactions**

These are as described below, and are the same as are present in a pure liquid, though we must always remember that they may well be **altered** or **modified** as a result of the presence of the solute.

The interactions present are:

(a) attractions between solvent molecules
(b) repulsions between solvent molecules.

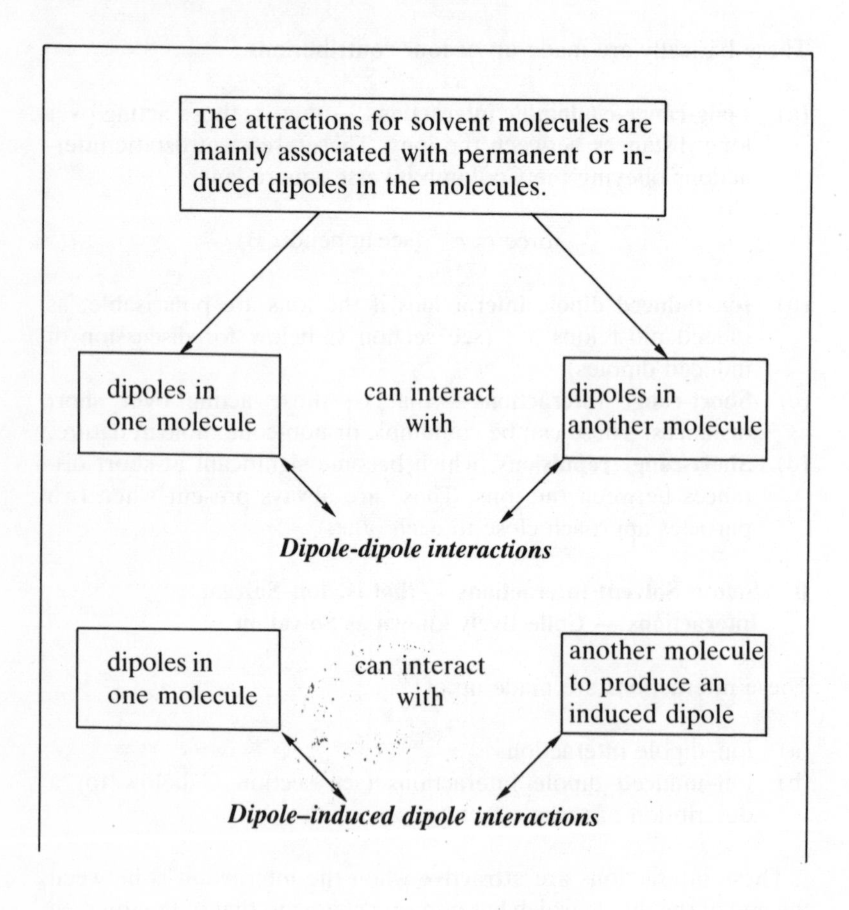

The attractions for solvent molecules are mainly associated with permanent or induced dipoles in the molecules.
dipoles in one molecule
can interact with
dipoles in another molecule
Dipole-dipole interactions
dipoles in one molecule
can interact with
another molecule to produce an induced dipole
Dipole–induced dipole interactions

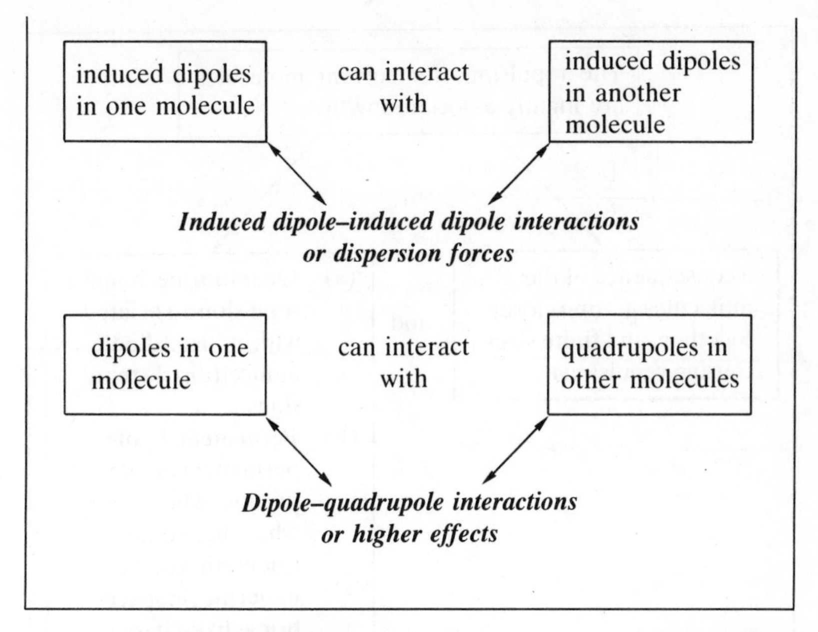
induced dipoles in one molecule
can interact with
induced dipoles in another molecule
Induced dipole–induced dipole interactions or dispersion forces
dipoles in one molecule
can interact with
quadrupoles in other molecules
Dipole–quadrupole interactions or higher effects

A quadrupole is a distribution of charge more complicated than a dipole. It can be exemplified as

$$\underline{\delta- \qquad \delta+} \qquad \underline{\delta+ \qquad \delta-}$$

The molecule CO_2 has a quadrupole of this nature:

$$\overset{\delta-}{O}=\overset{\delta+\delta+}{C}=\overset{\delta-}{O}$$

LIBRARY

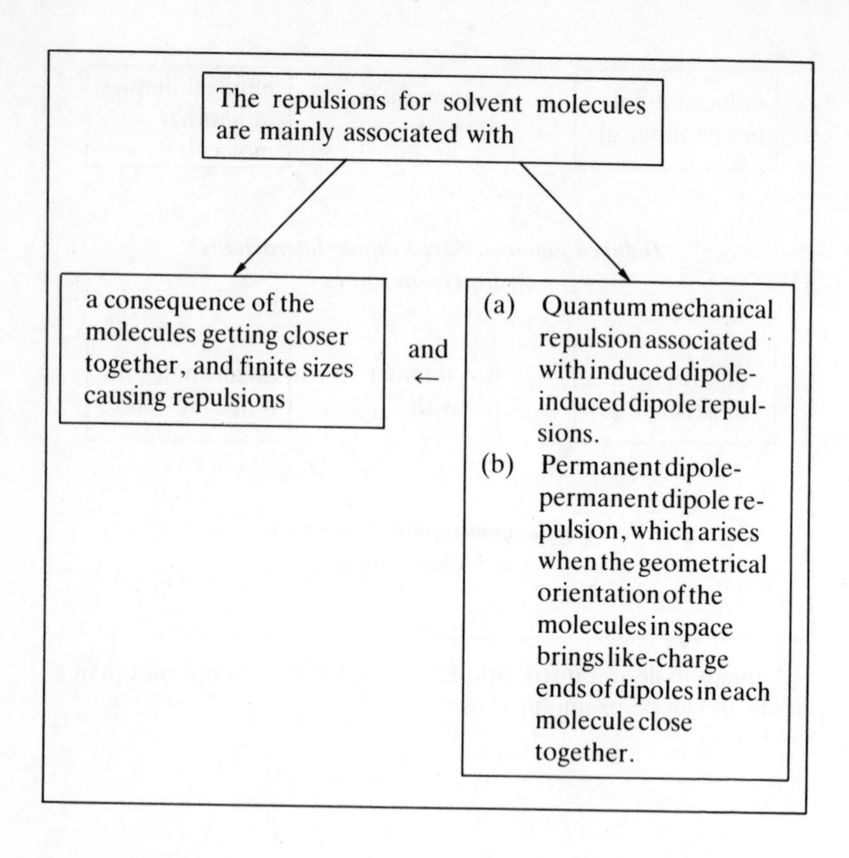

The repulsions for solvent molecules are mainly associated with
a consequence of the molecules getting closer together, and finite sizes causing repulsions
and ←
(a) Quantum mechanical repulsion associated with induced dipole-induced dipole repulsions.
(b) Permanent dipole-permanent dipole repulsion, which arises when the geometrical orientation of the molecules in space brings like-charge ends of dipoles in each molecule close together.

The Ideal Electrolyte Solution

In the ideal electrolyte solution all three interactions are present:

ion–ion
ion–solvent
solvent–solvent

} interactions

and the **ideal** electrolyte solution is defined as the infinitely dilute solution where the concentration of the solute (that is, the ions) $\rightarrow 0$.

Under these conditions:

1. Solvent–solvent interactions are significant, and are similar to those described in section C above.
2. Solute–solvent interactions are present, but are considered to be of less significance than solvent–solvent interactions because the ratio of solute to solvent is very low.
3. Solute–solute interactions are present, but are considered to be relatively small because of the low concentration of solute.
4. Repulsions are present, and will be significant for solvent–solvent interactions, but are considered negligible for solute–solute interactions because the concentration of solute → 0.

The Non-Ideal Electrolyte Solution

Physically this corresponds to all concentrations of solute (ions) other than that of infinite dilution where concentration of ions → 0. We can best look at what contributes to non-ideality by considering in turn what happens, as the concentration of solute increases, to the three types of interaction already discussed.

Under non-ideal conditions

1. Non-ideality corresponds to all **ion–ion** interactions which are **over and above** those considered to be present in the ideal solution, that is, we are considering any **modified** ion–ion interactions resulting from increase in the solute concentration. This is the **major factor** giving rise to non-ideality in electrolyte solutions. Here we must remember that there will be **three** ion–ion interactions:

   cation and anion
cation and cation } interactions
anion and anion

   There are a large number of modified interactions which can be considered as contributing to the non-ideality of the electrolyte solution. All of them result in increasing non-ideality as the solute concentration increases. They will be discussed in chapter 3.

2. Non-ideality corresponds to all **ion–solvent** interactions which are **over and above** those considered to be present in the ideal solution, that is, we are considering any **modified** ion–solvent interactions resulting from the increase in solute concentration. These interactions become more important at **high** concentration and the contribution of ion–solvent interactions to **non-ideality** in electrolyte solutions becomes important at **high** concentrations. An ion can interact with the solvent and can modify the solvent around it, or two ions could modify the solvent between them. This would correspond to an ion–solvent interaction different from the ideal case and would lead to non-ideality which would increase as the solute concentration increases. It would also result in modified solvent–solvent interactions as discussed below in 3.
3. Non-ideality corresponds to all **solvent–solvent** interactions which are **over and above** those considered to be present in the ideal solution, that is, we are considering **modified** solvent–solvent interactions resulting from the presence of the solute (cation + anion) at concentrations greater than infinite dilution. However this is a minor effect and of less importance than the contributions to non-ideality which are discussed in 1 and 2 above. For instance, the cation or anion, or both, could disturb the solvent structure present in the pure solvent, and this would in turn lead to modified solvent–solvent interactions which would then be considered as non-ideal. This modification would become increasingly greater as the solute concentration increases, giving rise to increasing non-ideality.

MACROSCOPIC MANIFESTATION OF NON-IDEALITY

The actual electrostatic potential energy of a real electrolyte solution is a sum of all these possible interactions. Each one makes its own contribution to the total electrostatic potential energy, and each contribution has a different weighting depending on the physical situation being considered. The **ideal** electrostatic potential energy is the sum of all interactions present in the ideal solution where the concentration of solute $\rightarrow 0$. The **difference** between the **real actual** electrostatic potential energy and the **ideal** value represents the **non-ideal** electrostatic potential energy. This can be shown to be equivalent to the **non-ideal** part of the total free energy, G, of the solution, and this non-ideal part of G is often termed the **excess free energy**, G^E, of the solution. This, in turn, can be considered in the simple primitive treatment of electrolyte solutions to manifest itself as an **activity coefficient** $f_{\pm}$ (sometimes written $\gamma_{\pm}$), or in a less primitive version as $f_{\pm}$ **plus a solvation term**. In modified treatments the excess free energy can be considered to manifest itself as

$f_{\pm}$ + ***solvation term*** + ***ion pair association term***

and in more advanced treatments the effects of shape, charge distribution, polarising power and polarisability will come in.

You will note that we are using the term **mean** activity coefficient, rather than an individual activity coefficient for the cation and one for the anion.

The activity coefficient is defined in terms of the activity and the concentration:

$$a = fc$$

Both activity and concentration are experimental quantities. We can **talk** about the activity and activity coefficient of each type of ion making up the electrolyte, but we **cannot** measure experimentally either the activity or activity coefficient of **individual** ions. Activity is a property of the electrolyte solution as a **whole**. We thus have to make do with **mean** activities and **mean** activity coefficients. The term mean is not used in its common sense of an average quantity, but is used in a rather different sense which reflects the number of ions which result from each given formula — that is

$$NaCl \rightarrow 2 \text{ ions}$$
$$CaCl_2 \rightarrow 3 \text{ ions}$$

For symmetrical electrolytes

$$f_{\pm}^2 = f_+ f_-$$

and for NaCl

$$f_{\pm}^2(\text{NaCl}) = f_{Na^+} f_{Cl^-}$$

For unsymmetrical electrolytes A_mB_n

$$f_{\pm}^{m+n} = f_{A^{n+}}^{m} f_{B^{m-}}^{n}$$

and for $CaCl_2$

$$f_{\pm CaCl_2}^{3} = f_{Ca^{2+}} f_{Cl^-}^{2}$$

2 Equilibria in Electrolyte Solutions

In a solution the solvent is the most abundant species. The solute can exist in various possible forms and we will confine ourselves to discussing the following only.

1. Free ions.
2. Undissociated molecules in equilibrium with free ions.
3. Ion pairs in equilibrium with free ions.
4. Complexes and chelates in equilibrium with free ions and other complexes and chelates.
5. Micelle clusters in equilibrium with free ions.

Situations 2–5 are equilibria and we can, in principle, formulate an equilibrium constant to describe each equilibrium.

DESCRIPTION OF THE EQUILIBRIUM

This means

(a) **postulating** which species are involved in the equilibrium, which implies **actually observing** the species to be present by chemical and physical means, or
(b) using **chemical judgement** to say that we are fairly certain that the species do exist.

Two examples will illustrate these points.

A. An Aqueous Solution of Ethanoic Acid

We postulate

the presence of undissociated acid molecules, ethanoate ions and hydrated protons all existing in equilibrium.

Evidence

(i) Ions are present, but not all of the ethanoic acid is present as ions. Conductance and colligative studies demonstrate both points.
(ii) Undissociated molecules are present. IR and Raman studies, and partitioning of the aqueous solution with an organic solvent, show this.

B. An Aqueous Solution of $CuSO_4$

We postulate

the presence of Cu^{2+}(aq), SO_4^{2-}(aq) and the $CuSO_4$(aq) ion pair.

Evidence

(i) Ions Cu^{2+}(aq) and SO_4^{2-}(aq) are present as shown by standard tests and spectroscopic evidence.
(ii) $CuSO_4$(aq) ion pair is present, and the evidence often comes from spectroscopic and ultrasonic studies.

Scientific methodology — a pause for thought.
It is no bad thing for us to sit down and assess just what chemical and physical evidence there is for saying that certain species are **actually present**.

It is also no bad thing to ask the question: are we just postulating their presence on **plausibility** grounds, or do we have **direct evidence**?

If you think seriously about this you may be amazed at how much of chemistry — and other scientific disciplines — is based on **plausibility and suggestion**, followed by the designing of experiments to test the 'truth' and justify 'the correctness of our formulation' in terms of a good fit between our ideas and experiment. When the fit is good, we often forget or do not realise that the existence of a good fit between theory and experiment does not actually demonstrate the correctness of our postulates. One eminent scientist called this "today's best guess."

In the present context this means that a good fit does not demonstrate that the species, for example the $CuSO_4$(aq) ion pair, is actually present. Unless we have **direct chemical evidence** of its existence, and, in reality, this is often very difficult to obtain — the existence is generally only of **inferential status**. All of this will become more evident when we study the 'status of the associated species'.

MEASURING THE EQUILIBRIUM CONSTANT: GENERAL CONSIDERATIONS

This is generally measured by comparing the behaviour expected for a given situation with what is actually observed. Examples will make this clearer.

A. A Weak Base

$$C_2H_5NH_2(aq) + H_2O(l) \rightleftharpoons C_2H_5\overset{+}{N}H_3(aq) + O\bar{H}(aq) \qquad (A)$$

1. If **all** the weak base is converted to free ions we could work out the behaviour **expected** for such a situation; for example, we would expect the conductance to behave in a certain way as the concentration altered.
2. We could measure the actual conductance for various concentrations, and find that the expected behaviour is **not** observed.
3. Because the **observed** behaviour is not what would be expected if all the ethylamine were in the form of free ions, we postulate the existence of process (A).
4. From the **quantitative** difference between the **observed** behaviour and the **expected** behaviour, we can calculate the equilibrium constant for process (A).

B. An Ion-Pair

$$Mg^{2+}(aq) + SO_4^{2-}(aq) \rightleftharpoons MgSO_4(aq) \qquad (B)$$

1. If **all** the $MgSO_4(s)$ which is dissolved in solution exists as free ions we can work out the depression of the freezing point of the solution **expected** for such a situation.
2. We can measure the freezing point depression and find that the expected value is **not** observed and that the freezing point is higher than expected.
3. To account for this we postulate that the number of particles in solution is reduced because of process (B) occurring.

4. From the **quantitative** difference between the **observed** and the **expected** behaviour, we can calculate the equilibrium constant for process (B).

All of this depends crucially on our measuring deviations from the expected or predicted behaviour for a solution containing **free ions only**.

If we cannot use methods designed to 'see' the ion-pair or complex directly, we need to use inference, and in such circumstances the fundamental requirements needed can be summarised as follows:

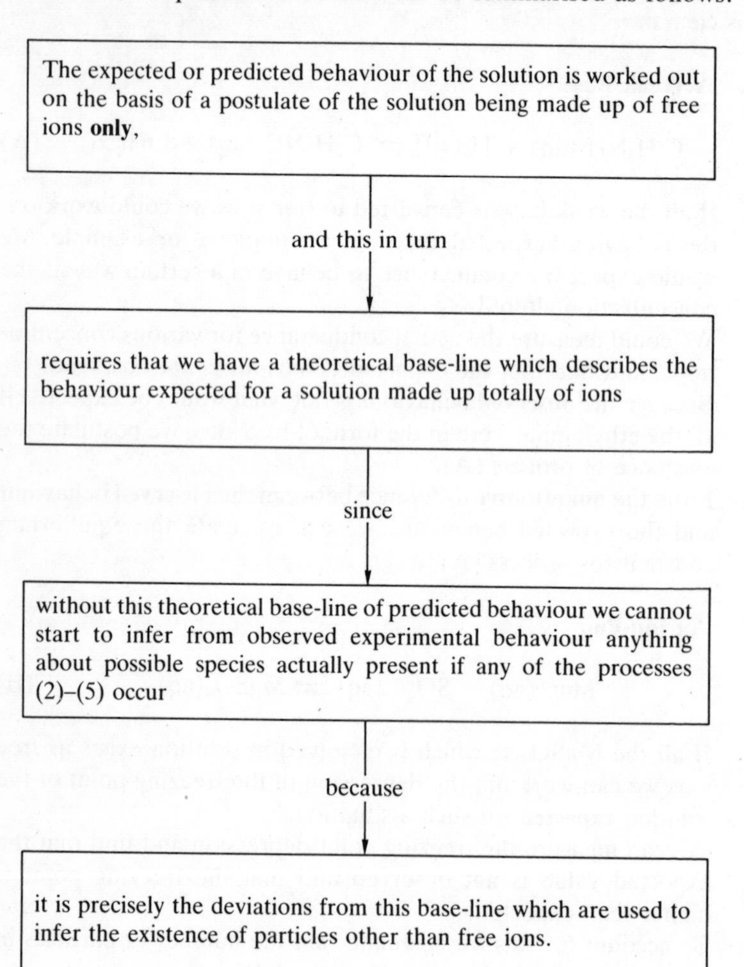

The expected or predicted behaviour of the solution is worked out on the basis of a postulate of the solution being made up of free ions only,
and this in turn
requires that we have a theoretical base-line which describes the behaviour expected for a solution made up totally of ions
since
without this theoretical base-line of predicted behaviour we cannot start to infer from observed experimental behaviour anything about possible species actually present if any of the processes (2)–(5) occur
because
it is precisely the deviations from this base-line which are used to infer the existence of particles other than free ions.

BASE-LINES FOR THEORETICAL PREDICTIONS ABOUT THE BEHAVIOUR EXPECTED FOR A SOLUTION CONSISTING OF FREE IONS ONLY

Setting up base-lines pre-supposes that we know **exactly** how a solution consisting of free ions only would behave over a range of concentrations, and this is where we have problems.

In chapters 1 and 3 we see that because the solute is made up of charged particles there are resultant interionic interactions which significantly affect the thermodynamic behaviour of the solutions. As the concentration tends to zero, these interionic interactions make a smaller contribution to the overall behaviour, and the solution approaches ideal behaviour. But as the concentration increases, the interionic interactions become more important, and significant deviations from ideality appear. These become greater at greater concentrations.

There are **THREE** major base-lines used in electrolyte studies:

1. The Debye–Hückel theory describing the dependence of the mean ionic activity coefficient for the solute on ionic strength (in effect, on concentration).
2. The Fuoss–Onsager extended conductance equation (or other conductance equations) describing the dependence of molar conductance on concentration.

These two base-lines are based on theories which are conceptually very similar, and deal directly with non-ideality resulting from coulombic interionic interactions.

The third base-line is totally different conceptually. It is based on:

3. Beer's law which describes the intensity of absorption of radiation by a solution as a function of concentration.

It does not deal with non-ideality, and assumes that non-ideality of the coulombic interionic interaction type has no effect on the theoretical expression.

BASE-LINE EQUATIONS FROM DEBYE–HÜCKEL AND FUOSS–ONSAGER

The equations are:

$$\log f_{\pm} = -A \mid z_1 z_2 \mid \sqrt{I} \quad (1)$$ Debye–Hückel limiting law

$$\log f_{\pm} = \frac{-A \mid z_1 z_2 \mid \sqrt{I}}{1 + B\mathring{a}\sqrt{I}} \quad (2)$$

$$\log f_{\pm} = \frac{-A \mid z_1 z_2 \mid \sqrt{I}}{1 + B\mathring{a}\sqrt{I}} + B'I \quad (3)$$ Debye–Hückel extended equation

$$\Lambda = \Lambda_0 - S\sqrt{c} \quad (4)$$ Fuoss–Onsager limiting law

$$\Lambda = \Lambda_0 - S\sqrt{c} + Ec \log c + Jc \quad (5)$$ Fuoss–Onsager extended equation

Constants (that is, A, B, S, E and J) are defined in chapter 3. Note that equations (1) and (4) have no adjustable parameters, but equations (2), (3) and (5) do.

It is important to remember that these equations describe the behaviour expected because of the interionic interactions between free ions — these being long-range interactions obeying the Coulomb law. No other deviations from ideality are included.

It is also important to remember that in:

equation 1

all quantities on the right-hand side are known or calculable in terms of known physical constants.

equation 2

contains one unknown parameter, $\mathring{a}$, whose magnitude we do not know *a priori*, nor is there any way of determining it unambiguously. To use equation 2 as a base-line, we have to assume a particular value of $\mathring{a}$ for the given solute.

equation 3

contains **two unknown parameters**, $\mathring{a}$ and B', whose magnitudes we do not know *a priori*. The value used for B' is even more difficult to estimate and there are fewer clues from other branches of chemistry to help us choose a physically realistic value. To use equation 3 as a base-line, we have to assume particular values for both $\mathring{a}$ and B' for a given solute.

equation 4

all quantities on the right-hand side are known or calculable.

equation 5

contains in the J term, one unknown parameter, $\mathring{a}$, and the comments on equation 3 are relevant here.

From these considerations the sensible approach might be to:

use equations 1 and 4 as base lines

These equations are rigorous, but only so at very low concentrations where it becomes very difficult to carry out precise and accurate experiments to test the validity of the equations. Nonetheless, their correctness has been demonstrated experimentally. This has required the development of very precise experimental techniques and apparatus, and has provided a tremendous impetus and challenge to the experimenter.

However, most experimental work is carried out at higher concentrations where the equations are certainly not valid, and we are therefore

forced to use equations 2, 3 and 5 as base-lines, and **forced** into coping with the problems introduced by the presence of unknown parameters.

BASE-LINE EQUATIONS IN SPECTROSCOPIC OR SPECTROPHOTOMETRIC METHODS

These methods are based on absorption of radiation by the species present in solution. We can describe the technique qualitatively by saying that:

> The position of the lines in the spectrum gives the nature of the absorbing species.
> The intensity of the lines in the spectrum gives the concentration of the absorbing species.

The intensity and the concentration are related by Beer's Law which is the base-line for electrolyte studies involved in spectroscopic techniques.

$$\log \frac{I_0}{I} = \epsilon c d$$

where
I_0 is the intensity of the incident radiation,
I is the intensity of radiation transmitted,
ϵ is a constant of proportionality (the molar absorption coefficient) which depends on the nature of the absorbing species,
c is the concentration of the absorbing species, and
d is the path length through which light is transmitted.

$\log(I_0/I)$ is measured directly by the instrument.

It is generally accepted that Beer's Law is valid over most experimental concentrations. Once ϵ is known for any absorbing species, then the expected absorbance, $\log(I_0/I)$, can be calculated for any stoichiometric concentration of that species. Deviations from the predicted values are interpreted as due to removal of the absorbing species by some equilibrium process, and a value of the equilibrium constant can be calculated.

Spectroscopic methods are also used to study equilibria in solution where **new peaks** are obtained when an equilibrium process is set up in solution. Observation of a new absorption is indicative of a new species being formed, and study of that new absorption using Beer's Law should generate an equilibrium constant.

MEASURING THE EQUILIBRIUM CONSTANT: SPECIFIC METHODS

1. Conductance Measurements

Deviations from the Fuoss–Onsager extended conductance equation are interpreted in terms of removal of ions by an equilibrium process, and from the magnitude of the deviations K_{assoc} is found.

Analysis of the data requires finding

$$\Lambda_0,\ \mathring{a},\ K_{assoc}$$

from the Fuoss–Onsager equation for associated electrolytes, (chapter 3), where for a process such as

$$Mg^{2+}(aq) + SO_4^{2-}(aq) \rightleftharpoons MgSO_4(aq)$$

$$K_{assoc} = \left(\frac{[MgSO_4]}{[Mg^{2+}][SO^{2-}{}_4]}\right)_{actual} \frac{1}{f_\pm{}^2}$$

where $f_\pm$ for the ion-pair is taken to be unity.

Finding Λ_0, $\mathring{a}$ and K_{assoc}

Λ_0 can often be found independently using Kohlrausch's Law and data from electrolytes known to be fully dissociated. In early work, Λ_0, $\mathring{a}$ and K_{assoc} were generally found by successive approximations, but most modern work uses computer analysis.

Two analytical procedures can be used:

(1) Obtain a statistical best fit by curve-fitting to the predicted behaviour to give **one** set of Λ_0, $\mathring{a}$ and K_{assoc}.

(2) Curve-fit to find the **best range of sets of** Λ_0, $\mathring{a}$ and K_{assoc}. Chemical judgement then decides which of these is chosen, but even if one set is chosen this is not irrevocable. Provided all the sets are given, other workers can still see whether they agree or not with the final choice.

Procedure (2) is by far the best, but unfortunately (1) is the procedure often used. Hopefully, with more and more workers using computer analysis, ranges of best-fit parameters will become the norm.

2. Activity Measurements

Some properties of a solution give a measure of the activity, a, rather than the concentration, c.

But these are related by the activity coefficient, f:

$$a = fc$$

where the activity coefficient approaches unity as the concentration $\to 0$.

The activity coefficient measures the deviation of concentration from activity, and can be calculated accurately from the two experimental quantities, a and c.

We measure the activity for a whole range of stoichiometric concentrations and calculate stoichiometric mean activity coefficients for each concentration. These are then compared with the expected value calculated from the Debye–Hückel equation for each experimental stoichiometric ionic strength. Any deviations are ascribed to removal of ions by an association equilibrium.

EMF studies, solubility measurements and freezing point measurements are standard ways of measuring the required stoichiometric mean activity coefficients.

3. Spectrophotometric Methods

Three major effects may be found:

1. A **new band** appears because of formation of an associated species, and analysis is easy.
2. A **change in intensity** in a band is attributed to formation of an associated species. Analysis is easy, provided the new associated species does not absorb around the same frequency.
3. **Overlapping bands** are found because of absorptions by one or other of the unassociated species and the associated species. Analysis is complex and requires curve-fitting procedures which are dependent on a knowledge of line shapes. As with the conductance curve-fitting, you must realise that there is often no one unambiguous best fit, and that all possible good fits must be looked at before a choice is made. Otherwise mistakes of interpretation are possible.

One advantage of the method is that the deviations observed from expected Beer's Law values are not attributable **even in part** to non-ideality, but are totally due to the setting up of an equilibrium process, in contrast to activity and conductance methods where deviations may be due to unaccounted-for non-ideality plus association.

4. Ultrasonics

A sound wave is passed through an electrolyte solution at equilibrium. This is done over a range of frequencies of the sound wave; and the absorption of sound energy by the solution, or the velocity of the sound wave through the solution, is measured for each frequency. The sound wave is equivalent to a pressure wave and passage of the sound wave will slightly disturb the equilibrium in the solution, and the solution has to return, or relax, to the required equilibrium position or positions. When it does so it absorbs energy from the sound wave, and this is shown by a change in the velocity of the sound wave or by a sudden increase in the absorption of sound by the solution. Each time this happens we know that an equilibrium process is adjusting itself. The **number** of times this happens tells us directly the **number** of equilibrium processes occurring in the solution which are being disturbed by the sound wave. Sometimes two or more peaks are superimposed, but these can be resolved by standard curve-fitting techniques (with reservations as mentioned earlier). In ultrasonics the decision as to how many equilibrium processes are involved is **direct** and not **inferential** as in other techniques. This gives it a tremendous head start over other methods.

However, identification of the process in a chemical sense is ambiguous as there is no direct chemical observation of the system. We can distinguish between chemical processes and physical processes, such as ion–solvent interactions or energy transfer, by the **position** of the maxima. Chemical equilibrium and the chemical species are inferred through a fit of theory plus inference with experiment, and the data may be susceptible to more than one interpretation. We have all the problems of data fitting and inference, but this is, as we have seen, a fairly common problem in all scientific studies.

5. Less Common Methods

These include freezing point, ion exchange, solvent extraction, polarography, Raman and NMR methods.

STATUS OF THE ASSOCIATED SPECIES

So far we have only postulated possible processes for removing ions from solution. What interests the chemist, and often the biologist, is knowing what species and equilibria do actually exist in the solution.

We must be clear on two points:

We must define precisely what we mean by the names describing the different types of associated species postulated.
We must ask ourselves whether we can actually decide between these different structures on the basis of experimental studies.

DEFINITIONS

Free Ions

Free ions are charged particles moving around independently on their own in solution with the only interactions with other charged species being long-range electrostatic interactions of the coulomb type. Solvent molecules often interact with ions to give a solvation sheath or shell.

Undissociated Molecule Formation from Free Ions

Our normal way of thinking about the ionisation of weak acids and bases is to regard such equilibria as being between undissociated molecules and ions. We rarely wonder whether they could be explained by an ion-pairing equilibrium. This is reflected even in the way we formulate the equilibrium.

$$CH_3C(=O){-}OH(aq) + H_2O(l) \rightleftharpoons CH_3C(=O){-}O^-(aq) + H_3^+O(aq)$$

We do not write it as

$$CH_3C(=O){-}O^-(aq) + H_3^+O(aq) \rightleftharpoons \underset{\text{neutral molecule}}{CH_3C(=O){-}OH(aq)} + H_2O(l)$$

or even as

$$CH_3\overset{\overset{O}{/\!/}}{C}-O^-(aq) + H_3^+O(aq) \rightleftharpoons \underset{\text{ion-pair}}{CH_3\overset{\overset{O}{/\!/}}{C}-O^-H^+(aq)} + H_2O(l)$$

Accepting that the covalently bonded neutral molecule is involved means that we have ruled out the possibility of the associated species being an ion-pair.

But, if we look at tables of dissociation constants of weak acids and bases, it is immediately clear that there is such a wide variation over many powers of ten that we can safely assume that the dominant effect is the chemical nature of the anion, for the weak acid, and the cation, for the weak base. We shall see later that when ion-pairs are formed the magnitude of the equilibrium constant is likely to be nearly independent of the chemical nature of the ions.

Ion-pairs, Complexes and Chelates

When we attempt to distinguish experimentally between the formation of

(a) ion-pairs from free ions

and

(b) complexes and chelates from free ions

we are in deep water, and considerable scepticism must be cast on the interpretation of many experiments which purport to make a clean-cut distinction. Also, different experimental methods may pick up behaviour typical of different types of associated species, so that a comparison of results from different methods may only add to the confusion. It is easy to give definitions, but much more difficult to decide what sort of species is present.

We shall take an unambiguous stance on the definitions, but also hope to make it abundantly clear that in many cases we will not be able to classify the results in any clear-cut manner into any specific category.

Ion-pairs from Free Ions

By an ion-pair we mean a physical entity with no specific intimate chemical interactions between the ions. The ions of the ion-pair move together as a single unit and are held together by electrostatic forces of the coulomb type acting over the short distances that the ions are apart in the ion-pair. These coulombic forces impose a certain degree of cohesion on the unit and this is sufficiently great to overcome the tendency for normal thermal motion to cause the ions to move around as separate independent particles each with its own translational degrees of freedom.

Because the forces holding the ions together are of this physical nature, they depend on three factors:

> The charges on the ions.
> The sizes of the ions; these are taken to imply the distance over which the forces act.
> The relative permittivity of the solvent in the vicinity of the ion.

It is very important to realise that these forces are therefore independent of the chemical nature of the ions.

We would, therefore, expect that electrolytes which have ions of the same charge and similar size would have equilibrium constants of similar magnitude **if** the associated species is an ion pair. This is found for some 2:2 sulphates, and for some cations of similar size which associate with Cl^- or I^-.

Charge Distribution on the Free Ion and the Ion-pair

The charge on an ion is usually unambiguous (for example, Mg^{2+} or NO_3^-) and the charge distribution for such ions is probably approximately spherically symmetrical. But the charge and the charge distribution for some ions may not be so clearly defined, for instance

$$CH_3CH_2CH_2CH_2NH_3^+$$

$$(CH_3)_3\overset{+}{N}CH_2CH_2CH_2C\begin{matrix}\nearrow O \\ \searrow O^-\end{matrix}$$

The charge on the ion-pair is the algebraic sum of the charges on the individual ions:

a +2 cation with a −2 anion gives an ion pair with an **overall** charge of zero,

while

a +3 cation with a −1 anion gives an ion-pair with an **overall** charge of +2

However, it is imperative to think more deeply than this. An ion-pair of zero overall charge must not be treated as though it were a neutral molecule. At best it can be regarded as a dipolar molecule, but it is probably more like a charge separated ion such as discussed in chapter 5.

$$Mg^{2+}(aq) + SO_4^{2-}(aq) \rightleftharpoons (\overrightarrow{MgSO_4})(aq)$$

a dipolar ion-pair

$$Mg^{2+}(aq) + SO_4^{2-}(aq) \rightleftharpoons (Mg^{2+}SO_4^{2-})(aq)$$

a charge-separated ion-pair

There are important implications arising from this:

(a) Interionic interactions between the ion-pair and free ions and other ion-pairs will be set up and contribute to the non-ideality of the solution.
(b) The ion-pair should be given a mean activity coefficient different from unity. This is rarely done.
(c) The ion-pair will probably be able to conduct a current, though its contribution will be small compared with that from the free ion. It is generally assumed to make a zero contribution.

When the ion-pair has an overall charge it should not be treated as though it were a **single** charge with a **spherically symmetrical distribution** of charge. Again it is probably more like a dipolar charged ion $(\overrightarrow{CaOH^{+}})(aq)$ or a charge-separated ion-pair $(Ca^{2}OH^{-})_{aq}$. For an overall charged ion-pair, we must always assign to it a mean ionic activity coefficient generally calculated from Debye–Hückel theory on the basis of the overall charge. We must also assign a non-zero molar conductance. In practice, its magnitude is very difficult to assess.

Size of an Ion and an Ion-pair in Solution

Most size correlations for free ions have used crystallographic radii which represent the bare ion. But there is no doubt that most ions are solvated in solution, though it is difficult to assess precisely the extent of solvation, and hence the size of the solvated ion.

Likewise, the ion-pair will be solvated, and we must get some estimate of its size. Furthermore, the change in solvation pattern on forming the ion-pair is of crucial importance.

Three limiting situations can be envisaged, but other intermediate situations are possible:

(1) An ion-pair is formed with no disruption of the individual solvation sheaths of the individual ions, so that in the ion-pair these solvation sheaths are in contact and solvent is present between the ions.

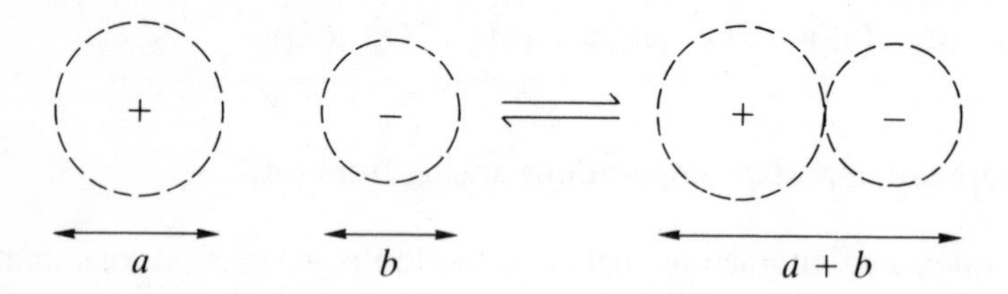

solvent spheres in contact

(2) An ion-pair is formed with total disruption of the individual solvation sheaths of the individual ions, so that in the ion-pair the bare ions are in contact and there is no solvent present between the ions.

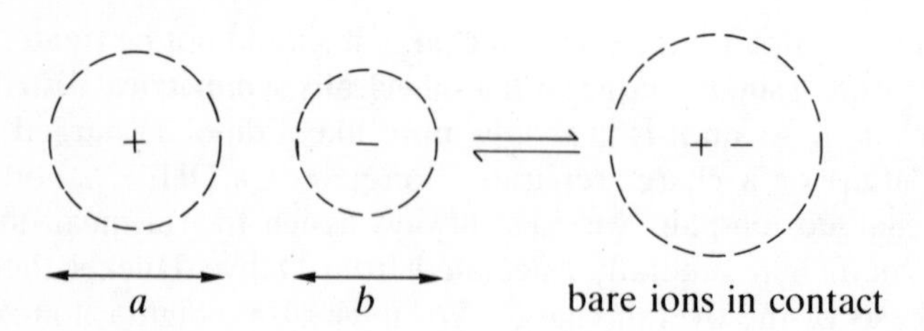

bare ions in contact

(3) An ion-pair is formed with partial disruption of the individual solvation sheaths of the individual ions, so that in the ion-pairs

some solvent has been squeezed out but there is still some solvent present between the ions.

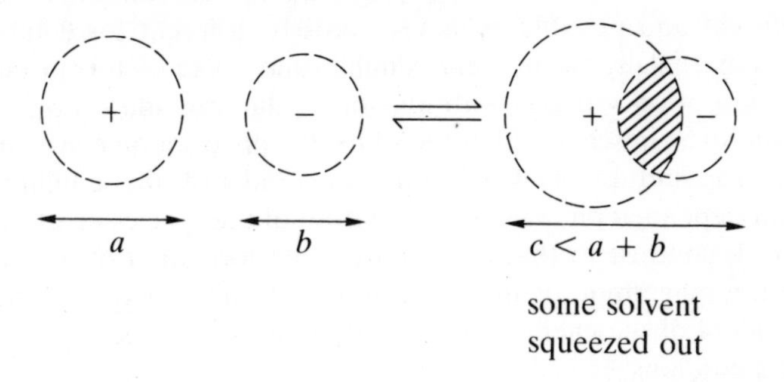

some solvent squeezed out

The ion-pairs which can be formed are thus not necessarily identical, and we must consider the possibility that different experimental methods may pick out and detect only one kind of ion-pair — for instance, detect contact ion-pairs but not solvent-separated ion-pairs.

A further formal definition can be proposed:

An **outer-sphere** ion-pair: is one where **one** or at most **two** solvent molecules lie between the ions.
An **inner-sphere** ion-pair: is one where the bare ions are in contact — all solvent sheaths have been eliminated from **between** the ions.

However, both inner-sphere and outer-sphere ion-pairs are still solvated as the **composite** unit — see the above diagrams.

Although our **definitions** can be quite unambiguous, **experimental** classification into inner-sphere and outer-sphere ion-pairs most certainly is not unambiguous, and may even, at best, be only a guess. This is exactly the same problem as will be encountered when discussing the formal and experimental distinctions between complexes and ion-pairs.

Complexes and Chelates from Free Ions

Where a complex is formed there is an intimate chemical interaction between the ions. Some electronic rearrangement is occurring and this results in covalent interactions, in contrast to the purely physical coulombic electrostatic interactions involved in the formation of an ion-pair.

If complexes and chelates involve intimate chemical interactions, we would expect that the extent of association should reflect the chemical nature of the ions involved. Equilibrium constants should be different and possibly even very grossly different for equilibria which superficially seem very similar and alike — for instance, association of one species with ions of similar size and charge. The situation is reminiscent of that found for the dissociation constants of weak acids and bases where the magnitude of the equilibrium constant depended on the chemical nature of the species involved. It is in stark contrast to that expected for the formation of ion-pairs, where the magnitude of the association constant is expected to be independent of the chemical nature of the ions involved.

The metal ions

$$Cu^{2+}, Ni^{2+}, Co^{2+}, Zn^{2+}, Mn^{2+}$$

have crystallographic radii which are very similar, and they all have the same charge +2. When they interact with oxalate

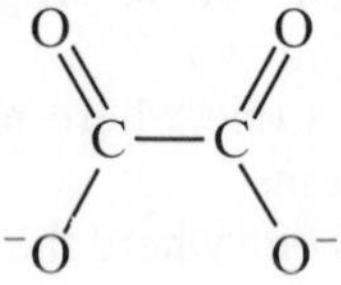

O O
C—C
$^{-}$O O^{-}

and with glycinate

O
NH_2CH_2C
O^{-}

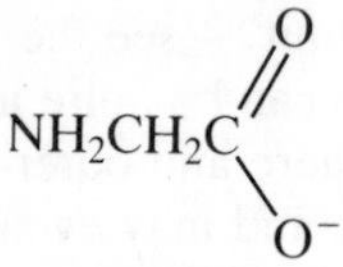

a wide variation in the association constants are found. The glycinates, for instance, have values ranging from 2.75×10^3 to 4.2×10^8 l mole^{-1}. Complexes are assumed to be formed.

On the other hand, it is believed that the interaction of the same metal ions with SO_4^{2-} results in an ion-pair. Here, the K_{assoc} values are very **similar**, ranging from 1.9×10^2 to 2.9×10^2 l mole^{-1}.

A thought for the perceptive student

Could these arguments be flawed because

(a) the sizes used are for bare ions, not solvated ones?
(b) the trends might be fortuitous since they represent a composite quantity, variation of $\Delta G^{\ominus}$ with size? Perhaps instead we should be correlating $\Delta H^{\ominus}$ and $\Delta S^{\ominus}$ with size, since

$$\Delta G^{\ominus} = -RT \log_e K$$

and

$$\Delta G^{\ominus} = \Delta H^{\ominus} - T\Delta S^{\ominus}$$

so that

$$\log K = \frac{-\Delta H^{\ominus}}{2.303RT} + \frac{\Delta S^{\ominus}}{2.303R}$$

Complexes formed from Ions and Uncharged Ligands

If an associated species is formed between an ion and an **uncharged** ligand, we usually assume that a complex is formed and that electronic rearrangements of a chemical nature have occurred.

When aqueous NH_3 is added to an aqueous solution containing Cu^{2+} ions an intense blue colouration indicative of electronic rearrangement occurs, and the main species formed is the complex

$$Cu^{2+}(aq) + 4NH_3(aq) \rightleftharpoons Cu(NH_3)_4^{2+}(aq)$$

Formation of an intimate chemical species implies a fairly drastic alteration of the solvation sphere around the Cu^{2+} (aq), with the NH_3 ligands displacing the solvent molecules from around the ion.

Chelates from Free Ions

If the ligand ion is a simple ion such as I^- or OH^-, then there is only one possible point of attachment irrespective of whether an ion pair or a complex is formed:

$$Pb^{2+}(aq) + I^-(aq) \rightleftharpoons PbI^+(aq)$$

If a ligand has two charges located in different parts of the ion, there are two possible points of attachment of the cation, and if the interactions are covalent the associated species is called a **chelate**.

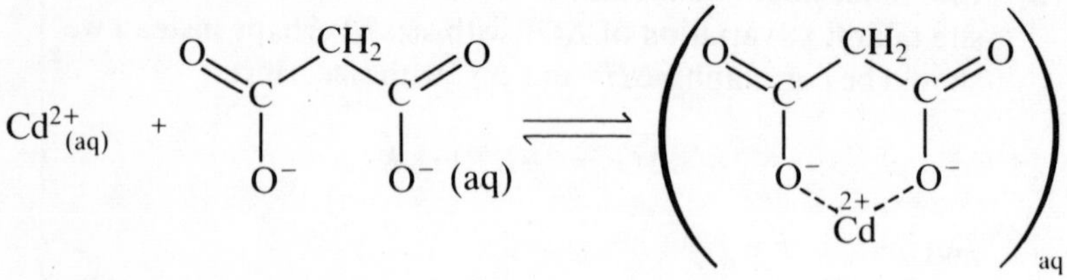

$$Cd^{2+}_{(aq)} + {}^-O(O{=})C{-}CH_2{-}C({=}O)O^-\,(aq) \rightleftharpoons \left(O{=}C(O^-){-}CH_2{-}C(O^-){=}O \cdots Cd^{2+} \right)_{aq}$$

Ligands can be ions but also have neutral points of attachment within the molecule — for instance, aminoacid anions such as glycinates. Does the metal ion then simply ion-pair with the carboxylate part, or does the amino part become involved to form a chelate? Is the equilibrium (A) or (B)?

$$\text{(A)} \quad Cu^{2+}\,(aq) + NH_2CH_2C({=}O)O^-(aq) \rightleftharpoons (NH_2CH_2C({=}O)O^- \quad Cu^{2+})_{aq}$$

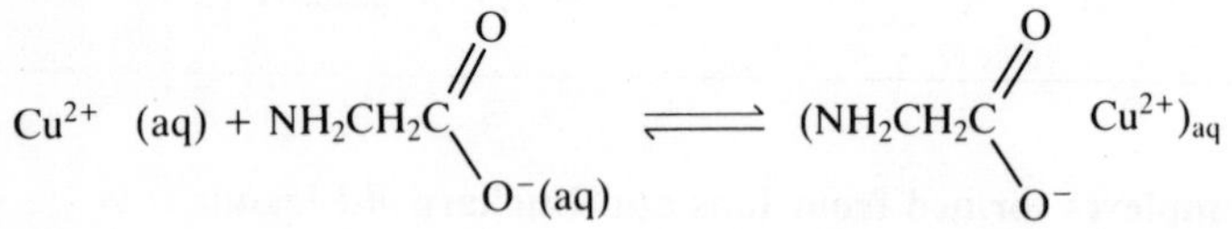

ion-pairing here

$$\text{(B)} \quad Cu^{2+}(aq) + NH_2CH_2C({=}O)O^-(aq) \rightleftharpoons \left(\begin{matrix} CH_2 - C{=}O \\ NH_2 \quad O^- \\ Cu^{2+} \end{matrix} \right)_{aq}$$

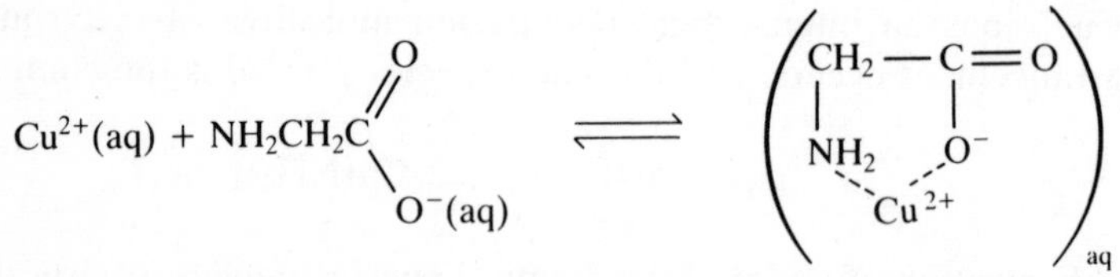

a chelate

Evidence suggests that the associated species is a chelate in this case.

Micelle Formation from Free Ions

Here clustering of ions of like charge occurs to give a cluster of colloidal size. Unambiguous detection of micelle formation is fairly easy experimentally because:

(a) Formation gives clusters of such a size as to be detected by standard techniques for colloidal solutions such as Tyndall's beam effects.
(b) Abrupt removal of such large numbers of ions from solution gives such a dramatic change in the properties of the solution that the effect is easily detected.

Electrolytes showing clustering properties are typified by paraffin chain salts where there is a long paraffin-like chain with a cationic group at the end, as in a quaternary ammonium group:

$$CH_3CH_2CH_2CH_2CH_2CH_2\overset{+}{N}(CH_3)_3$$

or a paraffin-like chain with an anionic group at the end like the carboxylates in soaps, and sulphonates in detergents:

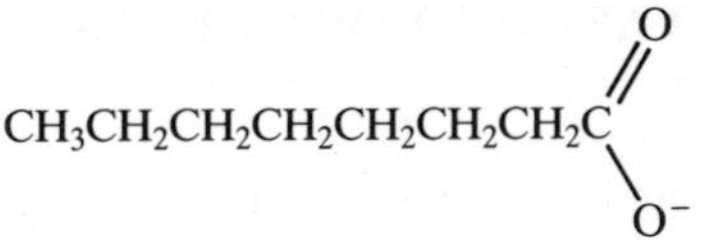

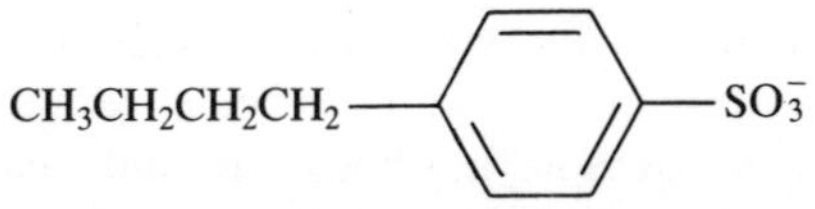

$$CH_3CH_2CH_2CH_2CH_2CH_2CH_2C\begin{matrix} // O \\ \backslash O^- \end{matrix}$$

$$CH_3CH_2CH_2CH_2-C_6H_4-SO_3^-$$

These electrolytes behave like a non-associated electrolyte up to a certain concentration, and then alter abruptly, with the properties changing dramatically. This is attributed to the rapid onset of micelle formation at a certain critical concentration. In the micelle, the paraffin chains face inwards with the charged groups lying on the outer surface where their charge is partially balanced by small simple ions of opposite charge, 'counter ions', fitting into the spaces of the cluster. Overall this has an effect on the properties of the solution similar to that expected if large numbers of ions are removed from solution.

Added salts can encourage the aggregation of ions to micelles, 'salting out'. This is of considerable importance in biological electrolytes, such as bile salts and phospholipids.

We must always distinguish clearly between micelles and polyelectrolytes. Many biological electrolytes are long chain species, for instance, proteins, where the individual molecule or electrolyte is similar in size to the micelle. Many such polyelectrolytes have positive and negative groups occurring alternately or irregularly along the chain. The biggest difference between micelles and biological polyelectrolytes is in the relative mobility of the polymer segments, which are often coiled but become uncoiled on addition of electrolytes which bind to the polyelectrolyte. Also, the micelle can break up into its individual ions — that is, it can come to pieces whereas the polymer does not. The electrostatic and configurational energy changes involved are not well understood and represent a challenge to the biologist and chemist alike. Solvation effects are also a major consideration, and have been shown in the last decade to be of crucial importance.

SPECIAL FEATURES OF SPECIFIC EXPERIMENTAL METHODS WHICH COULD LEAD TO DISTINCTIONS BETWEEN THE VARIOUS TYPES OF ASSOCIATED SPECIES BEING POSSIBLE

No technique has yet been devised which will enable an unambiguous and categorical distinction to be made between the various possible types of associated species. Claims have been made that certain techniques will detect specific types of associated species, but these should be viewed with scepticism and the evidence critically assessed. Ultrasonics and spectroscopic methods probably can allow us to distinguish with reasonable certainty between outer-sphere and inner-sphere ion-pairs, while conductance work and scattering of light experiments allows us to pick out micelle formation very easily, but these are quite exceptional achievements in the field of association.

1. Conductance and Activity Techniques

With the exception of micelle formation where there is a sudden large decrease in the effective number of ions present, and so a very abrupt and large decrease in conductance, there is nothing in these techniques which makes them inherently capable of drawing a distinction between undissociated molecules, ion-pairs and complexes. The two techniques simply measure the fact that charged particles have been removed from solution. They cannot in themselves tell us in what manner the ions are being removed.

2. Spectrophotometric and Spectroscopic Methods

This method is regarded as the one most capable of distinguishing between ion association and complexing.

Spectroscopic methods can appear to give a direct observation of the formation of a new species when a new absorption is found, or if a decrease in intensity of a cation absorption is found, on addition of an anion or ligand. But this does not tell us the **nature** of the new species.

Spectroscopic methods can also help us to distinguish between inner-sphere and outer-sphere associated species. For instance, there are theoretical and experimental reasons for assuming that:

(a) The visible spectrum picks up **inner-sphere** ion-pairs and complexes.
(b) The UV spectrum picks up **outer-sphere** ion-pairs and complexes.

3. Ultrasonic Methods

The beauty of the ultrasonic method is that it can often pick out how many equilibrium processes are present. If one peak in the absorption of sound/frequency graph is found, this tells us that **at least** one equilibrium is involved. Two peaks tell us that **at least** two equilibria are involved. Multistep equilibria can often be disentangled with relative ease. This is something which other methods are either incapable of, or which they can do only with difficulty and ambiguity.

However, there still remains the fundamental problem of identifying the process chemically. One clue which has been used is that the frequency at which absorption of the sound wave occurs is related to the rate constants for the processes involved. This means that fast processes are more likely to involve ionic interactions rather than covalent interactions. One other clue enables us to make distinctions between inner and outer sphere species. If the ***frequency of sound***

absorption is independent of the nature of the cation, an **outer-sphere** species is involved. When solvent molecules separate ions, this means the nature of the cation has less effect. If the ***frequency depends on the nature of the cation***, then an **inner-sphere** species is involved.

However, inference and chemical knowledge is still the main basis for attributing a particular type of chemical process to a given absorption.

OTHER MEANS OF TRYING TO IDENTIFY THE TYPE OF ASSOCIATED SPECIES INVOLVED

The following is only a brief survey with highly selective coverage, the main idea being to illustrate how we go about using experimental results to infer something about the chemical nature of the solution being studied.

1. A Combined Spectroscopic and K_{assoc} Magnitude Example

(i) The association constants for equilibria involving $Co^{2+}(aq)$ with $SO_4^{2-}(aq)$ and $S_2O_3^{2-}(aq)$ are very similar: $K_{CoSO_4} = 2.3 \times 10^2$ l mole^{-1}; $K_{CoS_2O_3} = 1.2 \times 10^2$ l mole^{-1}.

Inference. The associated species are of similar nature.

(ii) Both species show changes in the UV absorption, but formation of the thiosulphate associated species gives a change in the visible spectrum while formation of the sulphate does not.

Inference. $CoSO_4$ is an outer-sphere solvent-separated ion-pair; CoS_2O_3 is an inner-sphere contact ion-pair.

2. Ultrasonic Studies

Sulphates of Ni^{2+}, Mg^{2+}, Ca^{2+}, Mn^{2+}, Co^{2+}, Al^{3+}, Zn^{2+} and Be^{2+} give two absorption maxima, with the high-frequency maxima being independent of the nature of the cation, and the low-frequency maxima being dependent on the nature of the cation.

Inference. The following equilibria are set up:

(i) Two maxima mean two equilibria to be identified. A is ruled out because diffusion is unlikely to give absorption of sound at the frequencies involved.

$$M^{2+}(aq) + SO_4^{2-}(aq) \underset{A}{\overset{\text{diffusion}}{\rightleftharpoons}} (M^{2+}(H_2O)_2SO_4^{2-})aq$$

outer-sphere solvent-separated ion-pair

$$\underset{B}{\rightleftharpoons} M^{2+}(H_2O)SO_4^{2-})aq \underset{C}{\rightleftharpoons} (M^{2+}SO_4^{2-})aq$$

outer-sphere solvent-separated ion-pair — inner-sphere contact ion-pair

(ii) The high-frequency maxima which are independent of the nature of the cation suggest that both species are solvent separated. This suggests that equilibrium B is involved here.

(iii) The low-frequency maxima which are dependent on the nature of the cation suggest that the equilibrium involves a species with no solvent between the ions. This suggests that equilibrium C is involved here, since in this equilibrium the last H_2O molecule is being expelled to give a contact ion-pair.

3. Another Spectroscopic Example

When SO_4^{2-} is added to a solution containing $[(Co(NH_3)_5H_2O)]^{3+}$, a slow process is observed in the visible spectrum, and an immediate very rapid change is noted in the UV spectrum.

Inference. Two equilibria are involved:

A. $[Co(NH_3)_5H_2O]^{3+}(aq) + SO_4^{2-}(aq) \rightleftharpoons [Co(NH_3)_5SO_4]^{+}(aq) + H_2O$

B. $[Co(NH_3)_5H_2O]^{3+}(aq) + SO^{2-}(aq) \rightleftharpoons [Co(NH_3)_5H_2O^{3+} \text{ solvent } SO_4^{2-}]$

(i) The slow change in visible spectrum is attributed to equilibrium (A). A change in the visible spectrum suggests formation of a contact ion-pair $(Co(NH_3)_5^{3+}SO_4^{2-})aq$ or a complex ion $[Co(NH_3)_5SO_4]^{+}aq$. A slow change suggests formation of a complex with replacement of H_2O by SO_4^{2-}. This would mean an intimate chemical rearrangement involving covalent bonds.

(ii) The fast change in the UV spectrum is attributed to equilibrium (B). Change in the UV spectrum suggests an outer-sphere solvent-separated species. A fast change suggests a simple physical interaction such as formation of an ion-pair.

4. Use of Bjerrum's Theory

Bjerrum's theory (discussed in chapter 3) deals explicitly with formation of ion-pairs, and can be used to calculate an expected value for the association constant for an equilibrium between two ions. This predicted value can be compared with the observed value.

The conclusions are:

(a) If the two values are similar, then it is likely that ion pairing is involved since Bjerrum's theory deals explicitly with short-range coulombic interactions.

(b) If $K_{observed} \gg K_{Bjerrum}$, then it is likely that complexing should be considered since a much larger K_{assoc} would suggest additional interactions over and above those involved in Bjerrum-type association. The differences between the observed and predicted values would have to be sufficiently large for them not to be easily attributable to inadequacies of the theoretical base-line for analysing the data, or in the derivation of the Bjerrum equation itself.

5. Applications of Thermodynamic Reasoning

Most of these methods are based on correlations of the observed equilibrium constant with properties like atomic number of the cation, type of cation, type of anion, type of ligand, pK of the ligand if it is a weak acid or weak base, crystallographic radii, solvated radii, electrostatic interaction energies, ionisation potentials of the cation, symmetry of the ligand and other more exotic properties.

Ion association or complexing can be described as:

$$M^{n+}(aq) + A^{m-}(aq) \rightleftharpoons MA^{(n-m)+}(aq)$$

$$K_{assoc} = \left(\frac{[MA^{(n-m)+}]\, f_{MA}^{(n-m)+}}{[M^{n+}]\, f_{M}^{n+} [A^{m-}]\, f_{A}^{m-}} \right)_{eq}$$

$$\text{with } \Delta G^\ominus = -RT \log_e K_{assoc}$$

$$\text{Since } \Delta G^\ominus = \Delta H^\ominus - T\Delta S^\ominus$$

$$\log K_{assoc} = \frac{-\Delta H^\ominus}{2.303RT} + \frac{\Delta S^\ominus}{2.303R}$$

The last two equations show up the fundamental flaw in any such correlation. $\Delta G^\ominus$ and hence K_{assoc} are composite quantities, and values of $\Delta G^\ominus$ or K_{assoc} could be very similar for different systems **simply** because variations in $\Delta H^\ominus$ and $\Delta S^\ominus$ **compensate out**. We must be very careful that this is not the case. This is especially so when the conclusions drawn from experimental data depend crucially on the magnitude of $\Delta G^\ominus$ (or K_{assoc}), or on an observed variation of $\Delta G^\ominus$ or K_{assoc} with some properties such as those listed above.

This fundamental problem has only recently been more generally appreciated, and is one which has been shown to be of supreme importance when solvation is being considered. Ideally, we should measure both K_{assoc} and $\Delta H^\ominus_{assoc}$ and from this find $\Delta S^\ominus_{assoc}$.

Recently, a lot of work has been done to determine precision $\Delta H^\ominus_{assoc}$ values by calorimetry. This has greatly increased the number of situations where $\Delta H^\ominus$ and $\Delta S^\ominus$ can be found.

However unless **both** $\Delta G^\ominus_{assoc}$ (or K_{assoc}) and $\Delta H^\ominus_{assoc}$ have been measured, we are forced into using the old correlation procedures. Useful information can be obtained provided we do not push the interpretation too far.

6. Significance of the K_{assoc} values found

If association is slight, the data will have had to be collected at moderate or even relatively high concentrations — this is necessary to get a significant amount of association. Because the concentration is so high, we have to query:

(a) the adequacy of the Debye–Hückel model,
(b) the derivation of the Debye–Hückel theory,
(c) the derivation of the conductance theory.

We also have to ask whether the observed K_{assoc} actually measures something of chemical significance, or whether it is merely a parameter absorbing inadequacies of the models and the theories.

This **caution is general**, and is applicable to all methods of studying association in electrolytes. Self-consistency of the numerical analysis does not in itself argue for correctness of interpretation.

If association is significant, we can be more confident that we are measuring something which is physically equivalent to removal of ions from solution by some sort of association process. But, even so, the fact that we are forced into curve-fitting techniques limits our interpretation at the more detailed level. Exercises in fitting experimental data to predicted behaviour can only too easily cover up inadequacies of the theoretical treatments, and we must always be aware of this when we make chemical and physical interpretations of the parameters found.

3 Concepts in Electrolyte Theory

The Debye–Hückel theory which describes the effect of concentration on the activity coefficient of an electrolyte is central to all theoretical approaches to electrolyte solutions.

We shall be looking at what concepts are used in the theory at the **microscopic** level. We will also point out the ways in which they fit into the verbal framework of the theoretical study. No attempt will be made to develop the theories mathematically.

It is instructive to relate the concepts to:

(a) The **verbal** framework: in terms of the **physical** properties of the model(s) used.
(b) Thence to the **mathematical problems** this poses: what expressions are needed by (a)?
(c) Thence to the mathematical approximations needed because of mathematical complexity.

This then leads us to consider:

(d) How these approximations modify the model.
(e) The final mathematical solution relating macroscopic quantities to each other.
(f) Ways in which the model could be made more physically realistic.
(g) Recent advances incorporating these modifications.

The Debye–Hückel theory deals with **departures from ideality in electrolyte solutions**. The main experimental evidence for this non-ideality is that

(i) Concentration equilibrium constants are not constant,
(ii) Rate constants depend on concentration,
(iii) Molar conductances for strong electrolytes vary with concentration,

(iv) EMF expressions are not satisfactory when concentrations appear in the log [ ] terms,
(v) Freezing points of electrolyte solutions are different from what would be expected for ideal behaviour.

These departures from ideality become less as the concentration decreases, and behaviour approaches ideality. They are taken care of thermodynamically as the non-ideal part of the free energy — often called the 'excess free energy' G^E. This can alternatively be expressed in terms of activity coefficients.

Departures from ideality in electrolyte solutions have been shown (chapter 1) to be **mainly** due to

(a) electrostatic interactions (between the ions) which obey Coulomb's Law.

But we still ought to consider the smaller contributions from

(b) **modified** solvent–solvent interactions **over and above** those interactions present at infinite dilution, and
(c) **modified** ion-solvent interactions **over and above** those interactions present at infinite dilution.

In the **primitive** simple Debye–Hückel model attention is focused on (a), that is, on the coulombic interactions as the source of non-ideality. The **aim** of the Debye–Hückel theory is to calculate the mean activity coefficient for an electrolyte in **terms** of the **electrostatic interactions between the ions**.

All other interactions which lead to non-ideality, such as (b) and (c) above, are **not** considered and will have to be superimposed on the primitive model when we consider ways in which the simple treatment has to be modified to bring it more into line with experimental results.

The **problem** is to calculate from the primitive model

the electrostatic potential energy of all interactions in the non-ideal case.

One **mode of attack** is:

(a) Choose a given ion and work out the electrostatic interaction energy between it and all the other ions in turn — this requires a knowledge of the distribution of all the other ions around the central ion.

(b) Take each of the other ions in turn and make it a central ion, and proceed as in (a).

(c) Sum all the interaction energies, counting each pair once only.

This is very similar in essence to Milner's approach (1912). It corresponds to a very simple physical model but results in some very complex and well-nigh intractable mathematics. Consequently this approach was dropped, though, as we shall see later, it has been revised recently in the Monte Carlo computer simulation methods of solving the problems of electrolyte solutions.

FEATURES OF THE SIMPLE PRIMITIVE DEBYE–HÜCKEL MODEL

1. Strong electrolytes are completely dissociated into ions.
2. Random motion of these ions is not attained: electrostatic interactions impose some degree of order over random thermal motions.
3. Non-ideality is due to these electrostatic interactions between the ions. Only electrostatic interactions obeying Coulomb's inverse square law are considered.
4. Ions are considered to be spherically symmetrical unpolarisable charges. Spherical symmetry implies an ion-size, and this size represents a distance of closest approach within which no other ion or solvent molecule can approach. Unpolarisable means that the ion is a simple charge with no possibility of displacement of charge in the presence of an electric field imposed externally, or imposed by the presence of other ions (see chapter 1).
5. The solvent is considered to be a structureless continuous medium whose sole function is to allow the ions to exist as ions, and whose sole property is manifested in the bulk macroscopic value of the relative permittivity (see chapter 1). No microscopic structure is allowed for the solvent, which means that we do not consider any

    (a) specific ion–solvent interactions,
    (b) specific solvent–solvent interactions,
    (c) alignment of the dipoles of the solvent,

(d) polarisability of the solvent to give induced molecular dipoles,

(e) possibility of dielectric saturation.

For discussion of these points, see chapter 1.

6. No electrostriction is allowed (see chapter 1).
7. The most **important feature** of the Debye–Hückel model is that we consider each ion to have an **ionic atmosphere** associated with it. This is treated in the theory as though it were a smeared-out cloud of charge whose charge density varies continuously throughout the solution. Because the solution is overall electrically neutral, the charge on the central ion is balanced by the charge on the ionic atmosphere.

Points 1–7 are the main features of the primitive Debye–Hückel model. Other aspects of the Debye–Hückel theory are illustrated by the mode of approach which we will consider shortly.

But because of the extreme importance of understanding what we mean by the term ionic atmosphere, it is vital to explain this more fully.

PHYSICAL MEANING AND SIGNIFICANCE OF THE IONIC ATMOSPHERE

In all aspects of the Debye–Hückel theory we always consider the elctrolyte solution to be made up of:

a chosen central reference ion, called the j ion	and	all the other ions (cations and anions) which are present, called the i ions

The salient feature of the model is that

all these other i ions which-we would normally think of as a set of discrete charges	are to be replaced by	a smeared-out cloud of charge which varies continuously from point to point with the charge density being greater when the point considered is nearer to the central j ion.

This smeared out cloud of charge density is called

> the ionic atmosphere of the central j ion, and in it no discrete charges are allowed

Furthermore, we can choose any ion to be our central reference j ion and once

we have chosen a certain ion to be the reference central j ion	all the other ions present will constitute that central ion's ionic atmosphere

But we could pick out an ion from this ionic atmosphere and make it a central reference ion:

choose one of these ions to be the new central reference ion	and all the other ions including the central ion of the first reference system will now be the ionic atmosphere of the second reference system

Summary

I. Each central ion which corresponds to one framework or reference system is always part of the ionic atmosphere of each of all the other ions which could instead have been chosen to be the central reference ion.

II. Although point I above is a description in terms of discrete charges for the central reference ion, the ionic atmosphere is treated **as though** there were no discrete charges in it, and it behaved as a smeared-out charge cloud.

The reason and absolute necessity for replacing the set of discrete ions which make up the ionic atmosphere by a smeared-out charge cloud **only becomes obvious** when we start translating these **verbal statements** of the **model** into the **mathematics** of the theory.

NAIVETY OF THE DEBYE–HÜCKEL MODEL

A re-read of chapter 1 taken in conjunction with statements 1–7 makes us appreciate just how naive the Debye–Hückel primitive model is. But when we look at the complexities of fitting this very simple model into a mathematical framework, and the even greater complexities of solving the mathematics, we can understand how badly stuck we are with a model which is so physically unrealistic. Even today, much though we would like to incorporate modifications to the primitive model **into the theory**, this has proved mathematically almost impossible.

DEVELOPING A THEORY FROM A MODEL

Setting up a physical model is a relatively simple task, developing the theory is not.

When theories are given in textbooks or presented in original papers they always follow a beautifully logical, well-reasoned, step-by-step argument (both verbal and mathematical) giving the impression that this is the way the theoretician works. But this is a false impression. Theories are rarely developed in this almost automatic step-by-step deductive process and many blind alleys are often pursued. Also, the theoretician often starts by stating what he is hoping to attain, and works backwards looking at what arguments and types of mathematics and mathematical frameworks the final objective implies. **Working backwards** is often the clue as to what goes into the model and what goes into the **forward moving** arguments.

The real situation is more one of a combination of:

(a) Working forwards from a model.
(b) Working backwards from a desired end.
(c) Working sideways to find modifications to both the forward and backward thinking.
(d) Digressing into other branches of the subject, or into other scientific disciplines, and using models, tools and mathematical resources from these fields to help solve the problem in question.
(e) Knowing the answer from experiment so that the process is really one of rationalisation rather than prediction.

All in all, the conceptual and mental processes are very complex, involved, circuitous, sometimes highly intuitive, and involve divergent thinking rather than the coherent, deductive thinking which the final presentation would superficially imply.

No author of an original paper would be allowed to present his ideas other than concisely, logically and coherently. He certainly could not write:

> "Well I thought of this idea at this stage, and this led me to think of such and such which made me work backwards, and led me on to such and such; which then made me realise there could be a connection between these two points if **only** I could resolve such and such. After many false starts and digressions I finally 'twigged' the point by doing this and assuming that, and I eventually got my solution. I then derived the whole thing logically and coherently, and it **NOW** all seems straightforward and simple, and to follow logically step by step with a **considerable degree of inevitability** which, if the truth be told, only became apparent once I'd gone up all the blind alleys and landed on the right route."

The trouble with presenting theories as inevitable mathematical conclusions following inexorably from the initial premises and axioms is that

(i) For the student it often obscures **why** certain steps were taken and what were the **results** of the steps in terms of the ease of development.
(ii) It can leave the student feeling "But why **this**, and not **that**?"
(iii) It can lead to an **apparent** understanding of the theory which in actuality is only superficial: a fact which becomes obvious when students start to think for themselves about the theory rather than simply **following through** the theory.
(iv) It tends not to stress what each step in the argument corresponds to physically.
(v) It seldom points out the weaknesses in the argument.

The approach here will try to take a middle course and will focus on the physical situation and its correspondence with the mathematical formulation.

DERIVATION OF THE DEBYE–HÜCKEL THEORY FROM THE SIMPLE PRIMITIVE DEBYE–HÜCKEL MODEL

Step 1: Stating the Problem

The aim is to calculate the mean ionic activity coefficient from the non-ideal part of the free energy. This is done in terms of the potential energy of the coulombic interactions between the ion and its ionic atmosphere. These interactions give rise to non-ideality.

The electrostatic potential energy of interaction between the ion and its ionic atmosphere = $(z_j e)\psi_j^*$

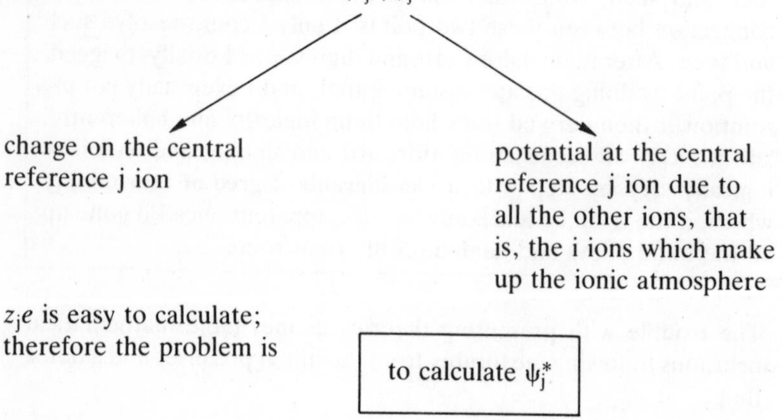

charge on the central reference j ion
potential at the central reference j ion due to all the other ions, that is, the i ions which make up the ionic atmosphere
$z_j e$ is easy to calculate; therefore the problem is
to calculate ψ_j^*

Step 2: The problem is

to calculate ψ_j^* in terms of other calculable potentials

(a) If the centre of the j ion is taken as the origin of the coordinate system used throughout the derivation, and if the diameter of the j ion is given by $å$, then ψ_j^* is the potential due to the ionic atmosphere at point $å$. We have talked about the potential ψ_j^* due to the ionic atmosphere at a point $å$ from the central j ion — that is, at a distance $å/2$ beyond the ion, since $å$ is the **diameter** of the ion. We have also talked about the potential ψ_j^* as being due to the ionic atmosphere at the surface of the ion. There is no mistake here because it can be shown from electrostatic theory that the potential due to the ionic atmosphere remains constant between the surface of the central j ion and a distance $å/2$ beyond the surface — that is, between $å/2$ and $å$ the potential due to the ionic atmosphere is constant and equal to ψ_j^*.

(b) If we could calculate the potential at **any** distance, r, from the j ion due to its ionic atmosphere, then ψ_j^* would just be a special case for the condition $r = å$.

(c) This focuses attention on potentials at distance r from the j ion. We can define

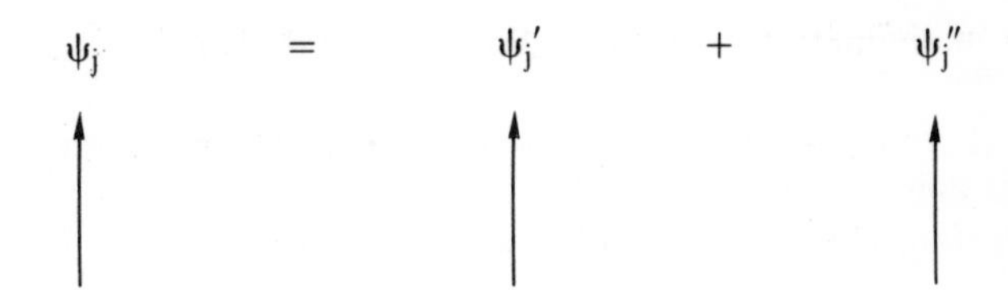

the total potential at a distance r from the central j ion due to the ion itself and its ionic atmosphere	the potential at a distance r from the j ion due to the ion itself	the potential at a distance r from the j ion due to the ionic atmosphere, that is, to all the other i ions

ψ_j is unknown but is calculable from Debye–Hückel theory.
ψ'_j is known and $= z_j e/4\pi\epsilon_0\epsilon_r r$ where ϵ_r = relative permittivity of the solvent.
ψ_j'' is unknown but can be found once ψ_j is calculated.
ψ_j^* is a special case of ψ_j'' for $r = å$.

Therefore the problem is

to calculate ψ_j

Step 3: The Problem is

Is there anything in physics, that is, in electrostatic theory, which could enable us to do this?

Fortunately the answer is yes, and is given in Poisson's equation. Poisson's equation relates the potential at any point to the charge density at that point for a continuous distribution of charge.

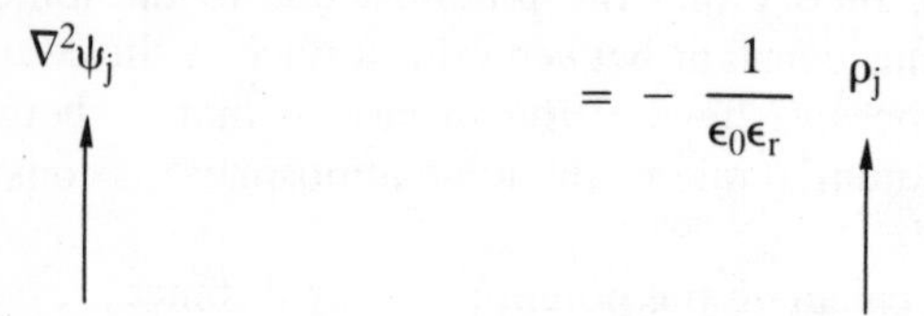

$$\nabla^2\psi_j = -\frac{1}{\epsilon_0\epsilon_r}\rho_j$$

This is the quantity we want

This relates to a **continuous distribution** of charge and is the **charge density** at any point.

∇^2 is a mathematical operator involving 2nd derivatives.

But there is no such thing as a charge density for discrete charges such as ions.

Therefore the next part of the problem is to

think of a physical process which would allow the discrete charges of the ionic atmosphere to be replaced by a continuous charge density.

Step 4: The Problem is to

replace discrete charges by a charge density.

We have talked in step 3 about **a distribution** of charge and in general about the **distribution** of ions in an electrolyte solution. This suggests bringing in **statistical mechanics** to help us, for example the Maxwell–Boltzmann distribution which for this situation gives us:

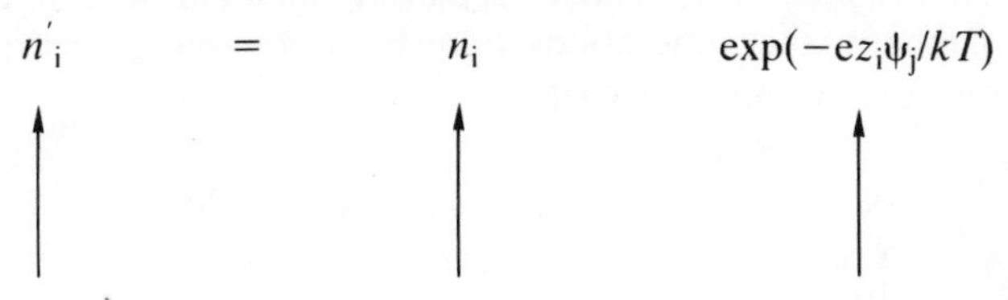

average local concentration of i ions in the region of solution considered	bulk concentration of i ions in the solution as a whole	ψ_j appears in Poisson equation and is the quantity we want

We have now got two equations, one involving ρ_j and the other n'_i:

Poisson's: $$\nabla^2\psi_j = -\frac{1}{\epsilon_0\epsilon_r}\rho_j$$ an equation in **two** unknowns ψ_j, ρ_j, hence cannot be solved

Maxwell–Boltzmann: $$n'_i = n_i\exp\left(\frac{-z_ie\psi_j}{kT}\right)$$ an equation in **two** unknowns ψ_j, n'_i, hence cannot be solved

Note that both these equations have ψ_j in them.

The problem now reduces to

> finding a relation between ρ_j and n'_i to enable Poisson's equation to be combined with the Maxwell–Boltzmann distribution.

Step 5: The Problem is

> to get some device which would substitute an average distribution of discrete charges, that is n'_i, by a continuous distribution of charge density ρ_j.

This step is the **crucial** aspect of the Debye–Hückel model. We have now found that it is **absolutely necessary** to introduce into the **model** the concept of smearing out all of the i ions into a continuous space charge whose charge density is assumed to be a function of the distance from the central j ion. This is the ionic atmosphere concept referred to previously. Having replaced our set of discrete charges of the i ions in the ionic atmosphere by a continuous distribution of charge, we can relate n_i' to ψ_j.

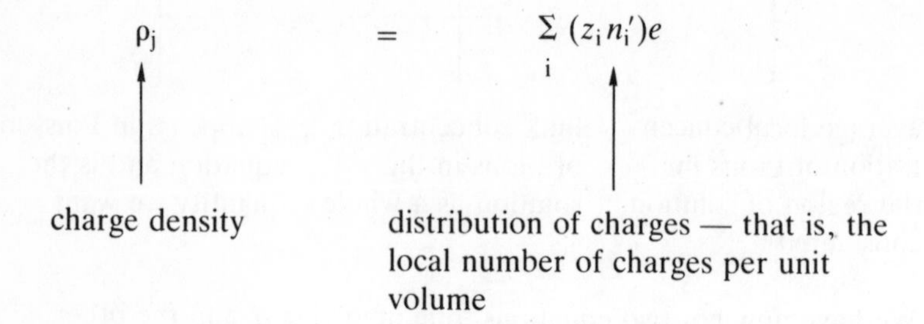

ρj = Σ (zi ni′)e
i
charge density
distribution of charges — that is, the local number of charges per unit volume

Substituting the Maxwell–Boltzmann expression for n_i' into this equation gives us

$$\rho_j = \sum n_i z_i e \exp\left(-\frac{e z_i \psi_j}{kT}\right)$$

and by substituting this value of ρ_j into Poisson's equation we then obtain an equation in one unknown which is thus in principle solvable. This new equation is called the **Poisson–Boltzmann equation:**

$$\nabla^2 \psi_j = -\frac{e}{\epsilon_0 \epsilon_r} \sum_i n_i z_i \exp\left(-\frac{e z_i \psi_j}{kT}\right)$$

Solving this equation would give ψ_j, from which ψ_j'' and then ψ_j^* could be found:

$$\psi'_j = \psi'_j + \psi''_j$$

now known — known — hence now known

ψ_j^* is a special value of ψ_j'' for $r = å$.

The problem has now reduced to:

solving the Poisson–Boltzmann equation, calculating ψ_j^* and evaluating the
electrostatic potential energy of interaction $= ez_j\psi_j^*$

Step 6: Solving the Poisson–Boltzmann Equation

This requires first that the exponential term is worked out, and the second-order differential equation is then solved.

Working out the exponential term can be done in two ways:

it can be done algebraically by expressing the exponential as a series and evaluating **all** the terms in the exponential. The calculations here are formidable and the approach was dropped.

But:
With the advent of computing techniques, Guggenheim (1957) was able to integrate the expression numerically to get an **exact** solution of the Poisson–Boltzmann equation. This major advance in the theory was made possible, not by an **advance in chemistry** but by an **advance in another field**, computer technology. Without computer techniques, exact solution of the Poisson–Boltzmann equation is well-nigh impossible.

it can be done by **approximation** by simple expansion of the exponential, and dropping out all terms other than the 1st and 2nd terms. It can be shown that for **symmetrical** electrolytes the 3rd, 5th and odd terms drop out automatically by cancellation, but do **not** do so for unsymmetrical electrolytes. So the theory is more accurate for symmetrical electrolytes than for unsymmetrical electrolytes, to the extent of including one more term, the 3rd. For both types of electrolytes, the 1st term drops out because of electrical neutrality and the exponential is thus **approximated** to one term only.

Expansion and approximation of the Poisson–Boltzmann equation to one term only gives:

$$\nabla^2\psi_j = \frac{1}{\epsilon_0\epsilon_r} \sum_i n_i z_i e\left(\frac{z_i e\psi_j}{kT}\right)$$

with consequent solution for the case of a spherical distribution of charge in the ionic atmosphere:

$$\psi_j = \frac{z_j e}{4\pi\epsilon_0\epsilon_r} \cdot \frac{\exp(\kappa \mathring{a})}{1 + \kappa \mathring{a}} \cdot \frac{\exp(-\kappa r)}{r}$$

where $\mathring{a}$ is the distance of closest approach.

κ is defined as:

$$\kappa^2 = \frac{2e^2N}{\epsilon_0\epsilon_r kT} I$$

where I = ionic strength

$$I = \frac{1}{2} \sum_i c_i z_i^2$$

$$= \frac{\frac{1}{2}\sum_i n_i z_i^2}{N}$$

Physical Significance of the Approximation Involved in Truncating the Exponential Terms to One Term Only, and in Limiting the Solution to a Spherical Distribution of Charge in the Ionic Atmosphere

It is very important indeed to realise just what these two approximations correspond to physically, and how they require a reappraisal of the physical basis of the primitive Debye–Hückel model.

1. *Truncation of the exponential*

This is only legitimate and valid for situations where

$$z_i e\psi_j \ll kT$$

which corresponds to

the electrostatic potential energy of coulombic interactions	$\ll$	thermal energy

and this can **only** hold for **very dilute** solutions where the ions are, on the average, at **large** distances from each other.

Now one of the **essential features** of the Debye–Hückel theory was that the

electrostatic potential energy of coulombic interactions	was comparable to	the average thermal energy represented by kT

and so we must unhappily accept that the Debye–Hückel theory can, at best, be only an approximate theory for normal working conditions, such as ionic strengths of 0.01 mole $litre^{-1}$ or above.

Summary

> The Debye–Hückel solution for non-ideality in electrolyte solutions is only valid at very low concentrations.

2. *Assumption of a spherical distribution of charge in the ionic atmosphere*

Any one given distribution of ions around the central j ion need not necessarily be spherically symmetrical, but if we **average** all possible arrangements this will correspond to spherical symmetry. A charge density necessarily corresponds to an **average** distribution of ions, so conversion of the Poisson–Boltzmann equation to spherical symmetry is purely formal.

But there is one very important limitation to this, and one of vital importance when considering large complex electrolytes such as found in aqueous solutions of biological materials. Here the central ion is non-spherical. An ion which is **not** spherically symmetrical, and many ions especially large ones certainly are not spherically symmetrical, may impose a **non-spherically symmetrical distribution** of charge around it, and this ought to be taken care of, **but is not**, in the theory.

Summary

The Debye–Hückel theory can only be approximate for non-spherical ions.

The next step in our problem is

to calculate ψ_j^* to evaluate $z_j e \psi_j^*$

Step 7

$$\psi_j \quad = \quad \psi'_j \quad + \quad \psi''_j$$

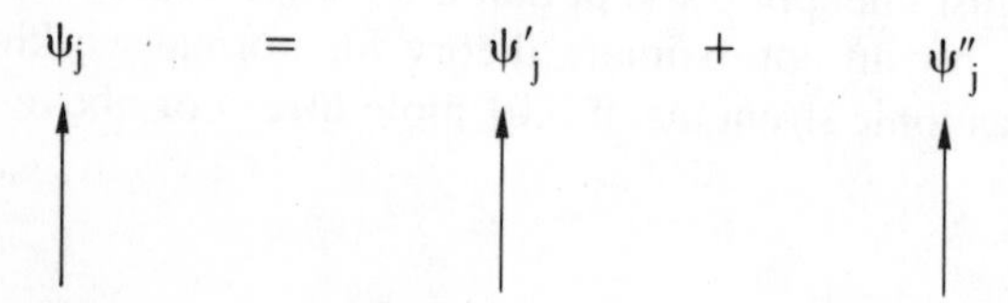

total potential at a distance r from the central j ion due to the ion itself and the ionic atmosphere: given by the approximate solution to the Poisson–Boltzmann equation	potential at a distance r from the central j ion due to the ion itself and equal to $\dfrac{z''_j}{4\pi\epsilon_0\epsilon_r r}$	potential at a distance r from the central j ion due to the ionic atmosphere: calculated from ψ_j and ψ'_j

$$\psi''_j = \frac{z_j e}{4\pi\epsilon_0\epsilon_r r}\left[\frac{\exp(\kappa å)}{1 + \kappa å}.\exp(-\kappa r) - 1\right]$$

and this holds for all r down to $r = \mathring{a}$. When $r = \mathring{a}$, ψ_j'' becomes ψ_j^*, and this reduces to

$$\psi_j^* = \frac{-z_j e}{4\pi\epsilon_0\epsilon_r} \cdot \frac{\kappa}{1 + \kappa\mathring{a}}$$

which is the potential at the surface of the j ion due to its ionic atmosphere. Therefore

the electrostatic potential energy of the coulombic interactions between the ion and its ionic atmosphere $= z_j e \psi_j^*$

$$= -\frac{z_j^2 e^2}{4\pi\epsilon_0\epsilon_r} \cdot \frac{\kappa}{1 + \kappa\mathring{a}}$$

and so the next step in the problem is

> to relate this electrostatic potential energy to the non-ideal part of the free energy and thence to the mean ionic activity coefficient.

Step 8: The problem is

> to calculate the mean ion activity coefficient, $f_{\pm}$

From the expression above it can be shown that:

$$\mu_{+}^{E} = \frac{-z_1^2 e^2}{8\pi\epsilon_0\epsilon_r} \cdot \frac{\kappa}{1 + \kappa\mathring{a}}$$

$$\mu_{-}^{E} = \frac{-z_2^2 e^2}{8\pi\epsilon_0\epsilon_r} \cdot \frac{\kappa}{1 + \kappa\mathring{a}}$$

where μ_{+}^{E} is the non-ideal part of the chemical potential for a cation of charge $z_1 e$, and is related to the non-ideal part of the free energy, and where μ_{-}^{E} is the corresponding quantity for an anion of charge $z_2 e$.

These two equations taken with the general expression

$$\mu_i^E = kT\log_e f_i$$

give

$$\log f_\pm = \frac{-A \mid z_1 z_2 \mid \sqrt{I}}{1 + B\mathring{a}\sqrt{I}}$$

where A and B are constants involving universal constants and $f_\pm^2 = f_+ f_-$ for symmetrical electrolytes (chapter 1).

The term '$\mathring{a}$' is somewhat arbitrary but is defined to be the value of r corresponding to the distance of closest approach of two ions — they cannot get closer than the sum of their respective radii. We are thus implicitly assuming '$\mathring{a}$' to be the same for both cation and anion. This is a quite considerable assumption, probably not valid for most electrolytes. This is especially so for large complex ions (such as found in biological situations) and for electrolytes which are not like simple ions such as Ca^{2+}, Na^+, OH^- or Cl^-.

When this equation is tested against experimental results it is very successful in accounting for behaviour at low concentrations, and we can be reasonably confident that the theory is basically correct at low concentrations. Having to test the theory rigorously at very low concentrations proved a great stimulus in developing precision techniques for deriving experimental values of $f_\pm$.

At moderate and higher concentrations deviations from theoretical behaviour become very apparent, and we shall be looking at ways of dealing with these problems later in the chapter.

Meanwhile, it is constructive to look again at the physical basis of the primitive Debye–Hückel model and its mathematical development to see where both could be modified and to consider whether this would be mathematically possible. What is written in chapter 1 on ions and solvent structure shows that the Debye–Hückel model is painfully naive and cannot even approach physical reality.

SHORTCOMINGS OF THE DEBYE–HÜCKEL MODEL

1. ***Strong electrolytes are completely dissociated***

As a result of feature (2) of the model (see below), consistent treatment should allow, in principle, the possibility that electrostatic interactions could be sufficiently large to result in some ion-pairing. The model must be modified to account for ion-pairing and this has been done by Bjerrum.

2. ***Random motion is not attained***

As mentioned above, the model has to be extended to take account of the **short-range** coulombic interactions of ion-pairs over and above **long-range** coulombic interactions of the ionic atmosphere dealt with in the theory.

3. ***Non-ideality results from coulombic interactions between ions***

This is not a sufficient description of what contributes to non-ideality. We must include many other types of interactions, for example:

short-range coulombic interactions
short-range quantum mechanical interactions
hard sphere repulsions
non-hard sphere repulsions
polarisability of ions — ion–induced-dipole interactions
induced-dipole–induced-dipole interactions
modified ion–solvent interactions
modified solvent–solvent interactions

Physically, all of these are simple ideas and are easy to discuss and explain, but unfortunately there are very considerable mathematical difficulties in incorporating them into the theory. Other than ion-pairing and very primitive solvation effects, nothing fundamental has been achieved for the others.

4. ***Ions are spherically symmetrical and unpolarisable***

Even some simple ions are not spherically symmetrical, and large ions, especially those encountered in solutions containing ions of biological importance, certainly are not. Likewise, many ions are not unpolarisable; even simple ones like I^- are highly polarisable, and the model needs to be grossly modified on both counts. But again this is so complex mathematically that not much has been done.

5. ***The solvent is a structureless dielectric***

This suggestion is manifestly untrue. The solvent consists of molecules which have structure. To cover more physically realistic situations the model needs considerable modification to include

ion–solvent interactions
solvent–solvent interactions
the effect of dielectric saturation

The complexity of the mathematics has meant that these features have not been incorporated into a new development of the theory, and suggests that a **new** model is needed.

6. ***Electrostriction is ignored***
Because of the intense fields near an ion, solvent molecules will get packed tighter together, but little consideration has been given to this.

7. ***Concept of a smeared-out charge density***
This is an absolutely crucial part of the model, but only spherical symmetry is assumed. And so for large non-spherical ions a modification is needed. This has not been done.

Shortcomings in the Mathematical Derivation of the Theory

These have been discussed as the theory was developed, and will only be summarised here.

a. ***Use of the Poisson equation in its spherically symmetrical form***
This is inadequate if the ions are non-spherical and a modified Poisson equation is needed.

b. ***Poisson–Boltzmann equation in its complete form***
This contradicts one of the basic laws of electrostatics, that is the linear superposition of fields. The Poisson equation is linear, but the Maxwell–Boltzmann equation is exponential — both should be linear to satisfy the above law. This is a fundamental problem to the theory and to any other theory which makes use of any such combination. In fact this is really one of the big problems in the theory.

c. ***Truncation of the exponential***
This is serious even though it removes the problem of (b) by having both sides of the combined equation linear. But it imposes a very **severe restriction** to the theory because the approximation requires

$$z_i e \psi_j \ll kT$$

For most real electrolyte solutions this is manifestly untrue, and means that the theory can only apply at **very low** concentrations. Extension to higher concentrations is required, but the only real advance here has been Guggenheim's numerical integration by computer.

d. ***Distance of closest approach, $\mathring{a}$***

We are in a predicament here as to what this actually means physically. There has been much argument, but no real major advance has been made theoretically. A serious consequence is that the theory assumes $\mathring{a}$ to be the same for all pairs of ions, and this obviously is not true.

MODIFICATIONS AND FURTHER DEVELOPMENTS OF THE THEORY

Precision experimental results showed that theory and experiment deviated from each other considerably at moderate concentrations, and grossly at high concentrations. The theory under these conditions is obviously inadequate, and the following is a short summary of attempts to overcome these problems.

(I) ***Extensions to higher concentrations***. Gronwall tried to extend the concentration range by including higher terms of the exponential. This was mathematically complex and did not do much to advance the theory.

(II) ***Extensions to higher concentrations***. Guggenheim's developments were much more successful. He used the **complete** Poisson–Boltzmann equation and solved it by numerical integration using computer technology — a development in the application of the theory which was impossible without computers. He includes all terms in his calculations. His results still contain the contradiction of the linear superposition of fields theory, but nevertheless his work was of considerable significance and showed:

(a) The Debye–Hückel treatment is correct and rigorous for very low concentrations.

(b) His treatment extended the range of the theory to ionic strengths of 0.1 molal; and shows quite unambiguously that the Debye–Hückel approximation is invalid, at other than very low concentrations, for charge types higher than 1–1.

(c) He proposed that a more accurate equation would be

$$\log f_{\pm} = -\frac{A \mid z_1 z_2 \mid \sqrt{I}}{1 + \sqrt{I}} + 2A \mid z_1 z_2 \mid I$$

and has shown that this equation describes the experimental results well.

(d) He has combined his accurate numerical solution by computer with Bjerrum's theory of ion-association and shown that a more complete and accurate picture of the behaviour of electrolyte solutions can be achieved in this manner. This is probably the best and most successful advance based on the Debye–Hückel **primitive model**.

(III) ***Computer simulation*** techniques probably represent a very useful theoretical approach. Higher-speed, higher-memory modern computers give a fairly realistic hope that a much more direct solution to the theory of electrolyte solutions may be forthcoming. Unfortunately, such techniques demand a vast amount of computing time and this is expensive.

Essentially, what is done is to choose a model for our electrolyte solution specifying what interactions are inherent in the model. For each type of interaction, the potential energy of interaction calculated pair wise can be formulated.

It is crucial to realise that the choice of model is the most important part of the technique, and this requires a considerable degree of insight into and understanding of what goes on at the microscopic level in an electrolyte solution. The computing part of computer simulation techniques basically makes use of an extremely useful and versatile tool which extends considerably the type of calculation able to be handled. This aspect of computer simulation must not be allowed to obscure the great need for thought in the setting up of the physical model. Nor should the requirements of the computing be allowed to cause a change in the model simply to make the computing problems simpler. The physical model should always remain the central issue.

The techniques then used are of two types:

(a) Monte Carlo calculations
(b) Molecular dynamics.

(a) *Monte Carlo calculations*

These are essentially an updated version of the Milner-type calculation outlined in chapter 1. The intractable computational difficulties which faced Milner can now be handled by the computer. The biggest problem is to first suggest a credible potential on which to base the Milner-type calculation.

The essence is to calculate all interactions between all ions and **average** them. More advanced models would include interactions between pairs of ions and solvent molecules and pairs of solvent molecules. The particles making up the solution are displaced from an initial random distribution by a small amount and the total energy calculated for this new distribution. The procedure is repeated taking only any displacements which led to a lowering of the total energy, since the states of lower energy contribute the most to all statistical averaged properties of the electrolyte solution at equilibrium. Doing this on a computer opens up almost limitless possibilities for our model, and since computing techniques are involved we can incorporate or discard whatever feature we like, and compare the results with each other, and with experiment.

For instance, Monte Carlo procedures start from a whole set of conceivable distributions, and:

(i) For each of our distributions we could include in our energy of interaction any number of possibilities of interaction. We could, for example, include some or all of the interactions discussed earlier as well as coulombic long-range interactions.
(ii) We could try to make our parameters for our ions as realistic as possible, for instance, include terms to account for size, shape, distribution of charge in the ion, charged-separated ions.
(iii) We could include all sorts of ion–solvent, solvent–solvent, and modified ion–solvent, modified solvent–solvent interactions.

Hopefully, if we could do all sorts of combinations of (i)–(iii) and compare each result one with the other, and with experiment, a much deeper and more accurate picture of what is happening at the microscopic level could emerge, and consequently a better theory of electrolyte solutions might be forthcoming.

(b) *Molecular dynamics*

Such studies are even more demanding of computing skill and computing time than Monte Carlo calculations.

For molecular dynamics the computer is used to describe how an assembly of molecules would behave over a period of time. A model is chosen and the resulting equations describing this are fed into the computer. Because of the interactions allowed in the model, rearrangements in the distribution of the particles in the solution follow inevitably. In molecular dynamics we can watch what rearrangements follow inevitably from the first description fed into the computer, and we can do this as a function of time. The simulation is open to sampling for all rearrangements which lead up to equilibrium, with sampling intervals of 10^{-15} s possible.

Molecular dynamics is open to all the possibilities listed in (i)–(iii) of the Monte Carlo calculations.

Obviously this is what could, in **principle**, be done, not what has actually been done. But there is a vast potential and excitement in computer simulation techniques; and they could be equally well applied to other areas of chemistry, physics and biology.

One **very important feature** of these techniques is that the limitations involved are not limitations **forced** on to the model or on to the theory. The limitations are **for once** not physical, chemical or biological, they are computer limitations:

(a) limitations on the ingenuity of the programmer
(b) limitations of speed and memory of the computer
(c) time and financial limitations.

(IV) ***Lattice calculations for concentrated solutions***. Some theoreticians believe we should approach the electrolyte solution theory by working downwards from the fused electrolyte to the highly concentrated solution, to the concentrated solution, and that, hopefully, this treatment will join up with the extension to the Debye–Hückel theory for dilute solutions given by the Guggenheim–Bjerrum approach.

This means that we start with the **known** short-range order of the crystal. We then use the known distribution function of the fused electrolyte and take it to be a good approximation to the distribution in solutions of very high concentration. As the concentration decreases, this function should approach and approximate to the Debye–Hückel, Guggenheim–Bjerrum approach. A considerable amount of work has been done, but so far nothing of great benefit to the experimentalist has emerged.

(V) ***Effects of ion-pairing***. Two main approaches have been made here: Bjerrum's in 1926 and Guggenheim's in 1957 onwards.

An essential feature of the Debye–Hückel model is that electrostatic interactions between ions impose some degree of order on the random thermal motion of the ions. It is thus possible that some interactions are so strong that two ions could move around together as an independent pair. This pair of ions would be counted as an ion pair if it survived long enough on a time scale of several collisions. This is more likely to happen in concentrated solutions where the ions are close together and the coulombic interaction is likely to be sufficiently large.

Bjerrum defined ion-pairs as:

ions which are within a certain distance of each other are PAIRED or ASSOCIATED

and

ions beyond this distance are UNPAIRED, FREE, or UNASSOCIATED.

Bjerrum considered that the cut-off between paired and unpaired ions occurs when the electrostatic potential energy of interaction equals $2kT$ and this assumption enables a critical distance, q, distinguishing paired from free ions to be defined. Ions which are **further** apart than the distance q are defined to be free or unassociated, ions which are **closer** than the distance q are defined to be paired or associated.

It is useful to see how the Bjerrum critical distance, q varies with charge type: see table 3.1.

Since simple ions have mean ionic diameters of the order of 4–6Å, it is evident that deviations from the Debye–Hückel approximation are likely to occur for charge types of 2–1 or greater. These deviations are then handled by Bjerrum's ion-pair concept. For 1–1 electrolytes, the Debye–Hückel approximations are generally valid, and for these electrolytes (provided $\mathring{a} \gg 3.57$ Å) we tend not to expect ion-pairing in water, and experimentally this is what is generally found. For large complex electrolytes, unless highly charged, we would expect no ion-pairing, since $\mathring{a}$, the distance of closest approach, for such electrolytes certainly will not be 3.57 Å — it will be much larger.

Table 3.1 Variation of the Bjerrum critical distance q, with charge type: for water at 25°C

Electrolyte type	*q (Å)*
1–1	3.57
2–1	7.14
3–1	10.81
2–2	14.08
3–2	21.42
3–3	32.04

Bjerrum's Theory of Ion Association

So far we have talked about what is meant by an ion-pair, but we have not related our arguments to an equilibrium constant defining ion-pairing. Bjerrum's treatment calculated explicitly a value for the association constant for the reaction:

$$M^{z+}{}_{(aq)} + A^{z-}{}_{(aq)} \rightleftharpoons MA^{(z_+ + z_-)}{}_{(aq)}$$

where

$$K_{assoc} = \frac{[MA^{(z_+ + z_-)}]}{[M^{z+}][A^{z-}]} \cdot \frac{f_{\pm MA}}{f_{\pm}^2}$$

If β is the fraction of the ions associated to ion pairs, then

$$K_{assoc} = \frac{\beta}{(1 - \beta)^2 c_{stoich}} \cdot \frac{f_{\pm MA}}{f_{\pm}^2}$$

where

β is the quantity calculated from Bjerrum's theory.

$f_{\pm}$ is the mean activity coefficient for the electrolyte ions which are unpaired, and is calculated from the Debye–Hückel expression.

$f_{\pm MA}$ is the mean activity coefficient for the ion-pair, and takes a value dependent on the type of electrolyte involved:

(a) If the electrolyte is symmetrical, then the ion-pair will be overall uncharged, though it will be a dipole. Its mean activity coefficient is then taken to be unity.

(b) If the electrolyte is unsymmetrical, the ion pair will be charged and the Debye–Hückel theory is used to calculate the activity coefficient.

The problem thus reduces to calculating β from the Bjerrum theory.

The calculation is based on the Debye–Hückel model and the following equation results, enabling β to be calculated:

$$\beta = 4\pi n_{\mathrm{i}} \int_{å}^{q} \exp\left(\frac{-z_{\mathrm{i}}z_{\mathrm{j}}e^2}{4\pi\epsilon_0\epsilon_{\mathrm{r}}kT_{\mathrm{r}}}\right) r^2 \mathrm{d}r$$

where

$$q = \frac{z_{\mathrm{i}}z_{\mathrm{j}}e^2}{8\pi\epsilon_0\epsilon_{\mathrm{r}}kT}$$

The integral has limits $å$ and q, since ions within the distance $å$ to q are defined as paired. Once β has been calculated, K_{assoc}, the theoretical association constant for the electrolyte, can be calculated.

Bjerrum's theory has been criticised because it involves an arbitrary cut-off at a critical distance q between ion-pairs and free ions. It is felt that a more realistic situation would be one which would allow more of a 'fall-off' between paired and free ions as the distance between them alters.

Nonetheless, despite this artificiality, the Bjerrum theory coupled with the Debye–Hückel theory, especially with Guggenheim's numerical integration, has proved a very useful and relatively successful tool in discussing electrolyte solutions.

There have been attempts to modify Bjerrum's treatment to remove this arbitrariness, but none has been used to any great extent in the interpretation of experimental data.

(VI) ***Empirical Extensions***. One final way of coping with observed deviations from the behaviour predicted by the Debye–Hückel expression is not to do any further theoretical calculations but to work at **empirical** extensions to the **limiting law** to take it to higher concentrations. These methods **assume** that the Debye–Hückel theory is **valid** at very low concentrations.

The empirical corrections can be summarised in three separate items:

(i) They add a term linear in I to extend the equation to higher concentrations.

(ii) They fit terms into the equations which allow for solvation of the ions.

(iii) They allow ion association and association constants to take account of all other deviations.

These points have all been dealt with in other chapters, and it would be worth while to recheck them now.

THEORIES OF ELECTROLYTIC CONDUCTANCE

The Debye–Hückel theory is a study of the **equilibrium** properties of electrolyte solutions, where departures from ideal behaviour are considered to be a result of coulombic interactions between ions in an **equilibrium** situation. And so we use equilibrium statistical mechanics to calculate an **equilibrium** Maxwell–Boltzmann distribution of ions.

When a current is passed through the solution, the current is being carried by the ions moving under the influence of an externally applied field. This is now a **non-equilibrium** situation, and is an inherently more difficult situation to handle theoretically.

All theories of electrolyte conduction use the Debye–Hückel model, but we now have to amplify that model to take into account extra features resulting from the movement of the ions in the solvent under the applied field. Ions move around at random anyway, but when a field is applied there is a strong tendency for the ions to move in one direction. In particular, we must consider what happens to the ionic atmosphere when the central reference ion moves under the influence of an applied external field.

As the ion moves, it normally must always have an ionic atmosphere associated with it. At each stage of movement, a new ionic atmosphere must be built up around the moving ion, and it takes time for this to happen. The ionic atmosphere has to **build up in front** of the moving ion and to **decay behind** the moving ion. Because this process cannot occur instantaneously, the net result is that the ionic atmosphere gets displaced with respect to the moving ion. There is more of the ionic atmosphere **behind** the ion than there is in **front** of the ion.

The ionic atmosphere is thus unsymmetrical, and the process giving rise to this asymmetry is called the **relaxation effect**.

The effect of the asymmetry of the ionic atmosphere when the ion is moving forward in the direction of the applied field or potential is to 'pull the ion back' so that the velocity of the ion in the forward direction is **less** than it would be if the ionic atmosphere were symmetrical around the moving ion. The ionic conductance of the ion is thus **less** than it would be if the ionic atmosphere were symmetrical.

We also have to consider the effect of the solvent on the **movement** of the ion and its ionic atmosphere under an applied field. This is discussed under the heading the **electrophoretic effect**.

When an ion moves, it carries its ionic atmosphere with it. This ionic atmosphere is made up, in part, of ions which have the same charge as the central ion and will move in the same direction as the central ion under an external field. It also contains ions which will have an opposite charge, and will move in the opposite direction.

If the central reference ion and an ion of opposite sign in the ionic atmosphere are passing each other, solvent will be pulled along with each ion but in opposite directions. So the central reference ion of the pair will, in effect, 'see' solvent streaming past itself in the opposite direction, and this will exert a viscous drag on the central reference ion, slowing it down.

If the central reference ion and an ion of like sign in the ionic atmosphere are passing each other, the differing speeds of the two ions will cause streaming of the solvent past the central reference ion. If the ion of the ionic atmosphere is overtaking the reference ion, the reference ion will be pulled along by the solvent streaming past it in the same direction, resulting in an increase in its velocity. If the central reference ion is overtaking the ion of the ionic atmosphere, it will experience a viscous drag by the solvent around the ionic atmosphere ion which will make this solvent appear as though it were, in effect, moving in the opposite direction to the central ion.

These effects are covered by the general term 'electrophoretic effect', and their **net** effect is always to **slow** the ion down, resulting in a **lower** ionic conductance than would be expected if there were no ionic atmosphere.

CONFIRMATION OF THE EXISTENCE OF THE IONIC ATMOSPHERE

There are two experiments which beautifully illustrate the correctness of the idea of an ionic atmosphere and its manifestation in terms of relaxation and electrophoresis when the ion moves under the influence of an external field. If an **alternating** potential (or field) of very high frequency (10^7 s^{-1}) is applied, we find that the molar conductance for an electrolyte solution is significantly higher than that for ordinary frequencies (50 s^{-1}). This effect (the Debye–Falkenhagen effect) is interpreted on the basis that at very high alternating frequencies the ion is impelled backwards and forwards so rapidly that the ionic atmosphere never has time to get into its asymmetric state. Hence the relaxation effect disappears and the velocity of the ions, and hence their individual ionic conductances, are much nearer what would be expected if there were no retarding effect of the ionic atmosphere.

In the other conclusive experiment **very large** potentials are applied, and the conductance values are found to be very much higher than under normal conditions. This implies that the ions are moving faster under the high potential conditions. This effect (the Wien effect) is a consequence of the extremely high velocities (about 1 m s^{-1}) which the ions achieve under such high potential gradients. These are so high that the ionic atmosphere is totally destroyed, and so there are no relaxation or electrophoretic effects associated with it.

These two experiments coupled with ordinary conductance and activity coefficient studies demonstrate conclusively the correctness of the basic postulate of the Debye–Hückel theory, that is, the existence of the ionic atmosphere. They also suggest that the main properties of the ionic atmosphere which must be taken into consideration in developing a theory of conductance are:

the relaxation effect (asymmetry of the ionic atmosphere), the electrophoretic effect (viscous drag on the ionic atmosphere).

EARLY CONDUCTANCE THEORY

If strong electrolytes are fully dissociated and there are no significant interactions between the ions, the molar conductance should be strictly constant over a range of concentration. But experimentally it is found that

$$\Lambda = \Lambda_0 - b\sqrt{c}$$

where Λ_0 is the limiting molar conductance found as $c \rightarrow 0$, as ideal conditions are aproached.

The deviations of Λ from Λ_0 are attributed to non-ideality associated with

(1) electrostatic interactions between ions,
(2) the relaxation effect,
(3) the electrophoretic effect.

Debye and Hückel tackled this non-equilibrium case once they had formulated their equation

$$\log f_{\pm} = \frac{-A \mid z_1 z_2 \mid \sqrt{I}}{1 + B\mathring{a}\sqrt{I}}$$

for the equilibrium situation in an electrolyte solution. They were closely followed by Onsager (in 1927) whose equation gave the basis of all modern theories of conductance. This was followed (in 1932) by a modified equation with Fuoss, but electrochemistry had to wait until 1957 before a full conductance equation of the level of refinement of the activity coefficient expression quoted above was derived.

There are very good reasons why there was this long delay of 30 years between formulation of the Debye–Hückel model, recognition of the necessity to consider effects of relaxation and electrophoresis, and formulation of a theoretical expression relating molar conductance with concentration.

The physical model is simple, but the mathematics involved are formidable and almost intractable. The problem does not lie in the numerical aspects of the calculation, but in the **analytical** development of the **types** of equations which are required. Computing technology will not help here, unless a model were developed to which molecular dynamics could be applied.

Because of the mathematical intractability of conductance theory, it is not easy, and perhaps impossible, to demonstrate in a short account the cross-linking between the model and the mathematics. And so only a brief summary of the results of conductance theory is given.

The early conductance theories given by Fuoss and Onsager in 1927 and 1932 used a model which assumed all the postulates of the Debye–Hückel theory which we have described earlier. The factors which have to be considered in addition are the effects of an asymmetric ionic atmosphere and viscous drag by the solvent on the movement of an ion under an applied external field. These effects result in a decreased ionic velocity and decreased ionic conductance and become greater as the concentration increases.

The theory aims to calculate the ionic velocities and thence the molar conductance as modified by relaxation and electrophoresis. This is done as a function of concentration.

In the 1932 derivation Fuoss and Onsager treated electrophoresis and relaxation independently. Other severe approximations were made in the treatments of the electrophoretic and relaxation effects. This was a very considerable approximation, but was done for mathematical ease.

The resultant expression is

$$\Lambda = \Lambda_0 - \left(\frac{a\Lambda_0 + b}{1 + \sqrt{c}}\right)\sqrt{c}$$

relating molar conductance with concentration

where a accounts for electrophoresis, and is a constant which involves the viscosity η, the hydrodynamic radius of the ion R, and the relative permittivity ϵ_r

and b accounts for relaxation, and is a constant which involves, among other things, the relative permittivity ϵ_r.

Theoretically this equation represents a **very primitive** approach, but this merely reflects the complexity of the mathematics which is involved in translating the statement of the physical model into the mathematical framework and derivation.

The 30 years of work which went into producing the second modified Fuoss–Onsager equation did not represent any alteration or improvement on the **model**, it merely allowed for a **few** less approximations to be made in the mathematical framework and derivation. However, these did represent a considerable advance on the physically unrealistic approximations of the earlier 1932 equation.

THE FINAL ALGEBRAIC FORMS OF THE 1957 EQUATION

For Unassociated Electrolytes

$$\Lambda = \Lambda_0 - Sc^{1/2} + Ec \log c + Jc - F\Lambda c + O(c^{3/2})$$

where $S = a\,\Lambda_0 + b$ as in the 1932 equation

$E = E_1\Lambda_0 - E_2$

$J = \sigma_1\Lambda_0 - \sigma_2$

F = constant $\times R^3$
O = a symbol meaning 'terms of the order of . . . '

and all concentration terms are **stoichiometric** concentrations.

S and E are constants defined by the same variables temperature, relative permittivity ϵ_r, viscosity η, charge type, and universal constants,
J is a parameter which is a function of the ion size,
F is a parameter correcting for viscosity effects and containing R the hydrodynamic radius.

This equation has been used with considerable success for unassociated electrolytes, in particular, unassociated 1–1 electrolytes, and the very few unassociated 2–2 electrolytes such as the disulphonates $SO^-{}_3CH_2SO^-{}_3$.

Deviations of observed behaviour from this equation have been ascribed to ion association, and the corresponding version for associated electrolytes has been derived.

Associated Electrolytes

$$\Lambda = \Lambda_0 - Sc^{1/2}{}_{act} + Ec_{act} \log c_{act} + Jc_{act} - K_{assoc}c_{act}f^2_{\pm}\Lambda - F\Lambda c_{act}$$

where concentrations are **actual** concentrations.

This equation has been used extensively to derive association constants from conductance work, and has been extremely successful in interpreting experimental data (chapter 2).

One very important aspect both of the Debye–Hückel theory and the Fuoss–Onsager equation has lain in the considerable impetus and stimulus which they have provided to the experimentalist who has striven to find more and more precise methods with which to test the theories. This has resulted in considerable improvements being made to conductance apparatus. It has also placed a very detailed emphasis on obtaining precision and accuracy of the measurements themselves. This has been of particular import when making measurements at very low concentrations where the experimental difficulties are greatest, but where it is important to test both theories in regions of concentration where we expect them to be valid. Such expectations have been vindicated by precision low concentration work and we can feel confident that both the Debye–Hückel equation and the 1957 conductance equation are accurate at low concentrations.

LIMITATIONS OF THIS TREATMENT

In many respects it is perverse and verging on the impertinent to criticise this equation when one considers the complexity of the treatment (a concise abbreviated version of the mathematical derivation of the 1957 equations runs to well over 100 pages), the enormous amount of mathematical skills required to solve the equations, and the massive expenditure of time and energy which was required.

However, despite the sophistication of the mathematics, it is necessary to state that this is only a very primitive treatment at the microscopic level.

The properties of the ions and solvent which are ignored are similar to those ignored in the Debye–Hückel treatment. These are really important properties at the microscopic level, but it would be a thankless task to try to incorporate them into the treatment used in the 1957 equation. Furthermore, Stokes' Law is used in the equation describing the movement of the ions. But this law applies to the motion of a **macroscopic** sphere through a **structureless continuous** medium. But the solvent is not structureless and use of Stokes' Law is approximate in the extreme. Likewise, the equations describing the motion also involve Einstein's equation relating to viscosity, and again viscosity is a macroscopic property of the solvent and does not include any of the important microscopic details of solvent structure.

It is quite obvious that further modification to conductance theory to take account of the shortcomings of the Debye–Hückel model as outlined would be even less fruitful than similar attempts on the Debye–Hückel theory itself.

SUMMING UP AND CONCLUDING REMARKS

Comparison of experimental results with theoretical prediction does not show the physical models of Debye–Hückel, Fuoss–Onsager, and Bjerrum to be grossly inadequate. The equations developed can handle the experimental data well and give good fits. The success of the Debye–Hückel, Fuoss–Onsager, and Bjerrum theories shows how often so many of the macroscopic properties of electrolyte solutions are insensitive to most aspects of microscopic detail, and this is why we can end up feeling that the theories are really more successful than they should be. This insensitivity of macroscopic properties to the details of the microscopic structure and behaviour is actually a fairly general phenomenon in physical chemistry. It arises through the relative importance and weighting of the various microscopic properties being fed into the theory. For example, in electrolyte theory the expressions end up insensitive to exclusion of microscopic details such as solvent structure and so forth, because the major contribution to the macroscopic quantities comes from the coulombic interactions between the ions — which is the one topic that the theory can handle relatively well. This results in the $c^{1/2}$ terms which appear in both Debye–Hückel and Fuoss–Onsager, and this low-order term (which thus takes high priority) **masks** the higher-order (low-priority) terms which would result from including microscopic details of ion–solvent and solvent–solvent interactions. These would result in higher-order terms than the leading $c^{1/2}$ terms and would, therefore, make a much smaller contribution to the observed macroscopic quantity. In general, this often happens in theories other than those for electrolytes, and this is particularly fortunate for the theoretician.

However, the model is shown to be inadequate, not by comparison of theory with experiment, but as a result of other experiments which go direct to the microscopic level, for instance ion–solvent and solvent–solvent interactions as studied by spectroscopic, ultrasonic and diffraction methods or computer simulations.

It is only when we think about the actual microscopic physical situation rather than data-fitting to a theoretical equation that the problems on inadequacy of the model show up.

The major theories of electrolyte solutions only make rudimentary allowance for solvation which other experiments such as the spectroscopic, diffraction and computer simulation studies show to be such an important aspect of the microscopic behaviour of electrolyte solutions. It is often solvation effects which result in fundamental differences in behaviour between solutions of different electrolytes (chapter 4). Yet all that comes out of Debye–Hückel, Fuoss–Onsager and Bjerrum is a distance of closest approach, $\mathring{a}$, whose numerical value is interpreted on rudimentary simple ideas of hydration or solvation numbers. No great wealth of microscopic detail on ion–solvent or solvent–solvent interactions is directly incorporated into the theories, and no great wealth of detail comes out. The fact that these equations fit the experimental data so well without having other than a rudimentary consideration of solvation is yet another example of the insensitivity of macroscopic relations to microscopic detail.

Perhaps a searching question for you to ponder, and with which to end a chapter which was intended to make you think about what goes on physically in the experiments which we conduct, and what goes on mentally when we interpret these experiments and develop theories to explain our interpretations, is the following.

Is it ever possible to know in advance, or even to know in retrospect, whether it is better to have

(a) an exact or very good solution to a rough and ready model, for example, Guggenheim's numerical integration

or

(b) a rough solution to a very descriptive, detailed and physically realistic model?

4 Solvation

Solvation occurs when a solute is dissolved in a solvent. We can classify solutes as:

1. ***Ions***: charged atoms or groups of atoms with positive or negative charges, or both in charge-separated ions.
2. ***Polar molecules***: uncharged neutral molecules with an overall dipole moment which may be the result of one individual polar bond within the molecule, or the result of a vector summation of several individual bond dipoles.
3. ***Non-polar molecules***: uncharged neutral molecules with a zero dipole moment. Vector summation of individual bond dipoles within the molecule may cancel to give a zero dipole moment. Or the molecule may contain bonds which are all completely non-polar, or virtually non-polar, so that the overall molecule is non-polar. This latter type of molecule is, in modern work, put into a separate category called apolar molecules.
4. ***Apolar molecules***: uncharged neutral molecules with an overall zero dipole moment and whose individual bonds are non-polar or virtually non-polar.
5. ***Biologically important molecules***: many of these are very large molecules which include both polymer and non-polymer molecules such as proteins and phospholipids. Depending on their environment they can be uncharged or charged, and they certainly contain regions which are decidedly polar and regions which are apolar. This particular aspect of biologically important molecules is often associated with their overall structure and biological function, and we will be looking at this in more detail later.

We can also classify **solvents** as polar, non-polar and apolar. But the more general classification uses the relative permittivity. There are high relative permittivity solvents such as (at 25°C):

water (78.30) HF (83.6) H_2SO_4 (101)
HCN (106.8) formamide (109.5)

However, the other more polar common solvents have much lower relative permittivities:

acetamide (59 at 83°C) methanol (32.6 at 25°C)
hydrazine (25 at 25°C) ammonia (22 at −34°C)

Non-polar liquids have relative permittivities of around 2, such as (at 25°C):

benzene (2.27) dioxan (2.21)

SOLVENT STRUCTURE

The structure of a crystalline solid is an extensive repeating fixed and regular pattern in three dimensions. A liquid, however, is characterised by random Brownian motion with no large-scale repeating periodic permanent structure throughout the liquid. Nonetheless, liquids do have a certain degree of transient structure. If we chose one molecule to be a central reference molecule there is a regular arrangement, rather than a random arrangement of other molecules around it. This arrangement is not on the large scale and is constantly being broken down and re-formed, but its lifetime is long on a Brownian motion scale. Structure in this sense **cannot** be defined by a set of unchanging atomic coordinates as in a crystal, but X-ray and neutron diffraction show conclusively that diffraction patterns characteristic of a repeating short-range structure **are** obtained from liquids. The processing of the data for liquids is necessarily different and more difficult than for solids, and what comes out of the experiments is an expression which tells us the chance of finding an atom or molecule, i, at a distance, r, (over 360°) from the central reference atom or molecule j. It is probably one of the most important, basic and characteristic properties of a liquid, and it tells us a lot about the **arrangements** of the molecules in a liquid.

With modern techniques we can approach this expression from two directions — experimental and theoretical.

1. ***The experimental approach*** uses X-ray and neutron diffraction. Neutron diffraction has contributed enormously to our

knowledge of the molecular structure of liquids. This is because it can pin-point light atoms such as hydrogen more accurately than X-rays can. However, modern technological advances, including synchroton-generated X-rays, give vastly superior X-ray data leading to greater scope for X-ray diffraction techniques in solution.

Knowledge of the pure solvent structure is necessary before we can hope to study the way in which the introduction of a solute molecule into a solvent can alter or disrupt the solvent's molecular structure and behaviour.

The diagram below is a **qualitative** representation of what comes out of a diffraction experiment.

A function which gives us information about the arrangement of molecules in a liquid

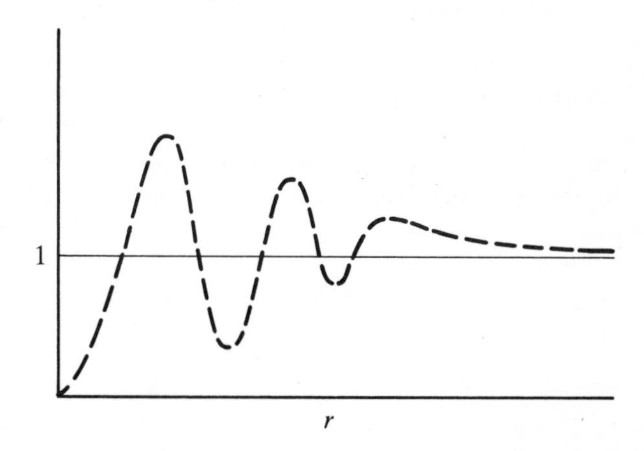

A peak tells us that there is a good chance of finding a molecule at a distance from the central molecule given by the position on the r-axis. The successive maxima tell us that there is structure in the liquid, and that the central molecule 'sees' another molecule at distances given by the peaks. When the peaks die away the central molecule no longer 'sees' a definite arrangement of molecules around it, and thereafter the arrangement appears random. The structure above is three shells of solvent molecules deep around the central molecule. The **arrangement** of the other molecules around the central molecule can be inferred — for instance, it could be tetrahedral or octahedral.

2. ***The theoretical approach***. With the advent of the modern fast computer and the development of Monte Carlo and molecular

dynamics techniques, it is now possible to deduce the information given in 1 from the simulation experiments based on assumed theoretical molecular potential energy functions. This is a development of supreme importance, well advanced for liquids and progressing rapidly for solutions. It is one of the main spearheads of the modern attack on molecular interactions in liquids and solutions.

Computer simulations are today an exciting advance for electrolyte and small solute studies. In the future they could well be an exciting and fruitful development for biological molecules and could possibly enable detail at the molecular level to be correlated with details of the structure of the macromolecule in solution and its biological function.

STRUCTURE OF LIQUID WATER

A 'partial structure' exists in water as a result of H-bonding between water molecules. The charge distribution in the H-bond is

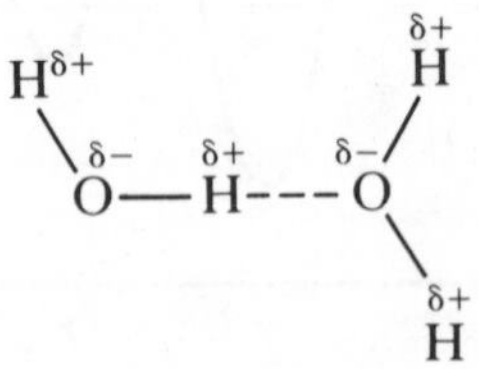

```
                      δ+
   Hδ+                H
    \                /
     \ δ−    δ+   δ−/
      O — H - - - O
                    \
                     \
                      δ+
                      H
```

and the most stable form is believed to be the linear H-bond. The H-bonding extends throughout the liquid but this does not mean that the H-bonded structure is fixed and static. Rather it is a dynamic structure with H-bonds constantly being broken and new ones formed. The structure, and on the average each H-bond, exists for a relatively long time of the order of 3×10^{-12} s. This is a longer time than that of Brownian motion, period of X-rays or IR radiation, which is of times shorter than 1×10^{-12} s.

The linear H-bond imposes a very open tetrahedral structure on the water molecules throughout the liquid. Diffraction experiments show **each** water molecule to be tetrahedrally surrounded by about **four** other water molecules at a distance of 0.28 nm with a next layer at 0.485 nm and a third layer at 0.58 nm. At distances further than this the central molecule no longer 'sees' the H-bonded arrangement as such, but sees only an apparent random distribution. Nonetheless, this is totally compatible with an extended H-bonded array throughout the liquid since each central molecule is part of the 'structure' around another molecule which could instead have been chosen as the central reference molecule.

The emphasis in this book is on aqueous solutions and we will not be going into details about the structure of other solvents. What you must realise is that the same approaches can be used to elucidate the structure of any solvent, but unfortunately nothing like the same detail or amount of data is available for other solvents, and even less for mixed solvents. Water is such an important solvent in chemistry, and is almost the exclusive one in biological systems, that the emphasis has been very much slanted to the study of water.

SOLUTIONS

In the past the emphasis of solvation has been on what happens to the **solute** when it changes from the pure state to the dissolved state in the solvent. Obviously the chemist has been very much aware of the role of the solvent in solvation, but the focus has always been on the **solvated solute species**.

More recently chemists have become aware that much more attention should be given to the **change** in solvent 'structure' from the pure solvent to that involved in the solution around the solute. This aspect of solvation could also be important to the biologist.

The structure of biological polymers and macro-molecules in solution, and their biological function, is crucially linked with the change in solvent structure around the solute when these substances are dissolved in solution. Without a knowledge of the details of this change in solvent structure, the biologist will be hampered in tackling the complexity of the aqueous solution biochemistry of biological materials, or in acquiring a full understanding of their biological function. Since studies on simple molecules are easier and can point the way for studies on more complex molecules, it is vitally important for biologists to have a good grounding in the chemistry of solvation of simple molecules and then of more complex molecules. It is also useful to see the respects in which these are comparable and different. Because biologically active molecules can be charged, an understanding of the current state of electrolyte theory and studies is also important.

Both chemists and biologists need to know something about the experimental and theoretical techniques for studying solvation. Experimental techniques include direct methods such as spectroscopic methods, diffraction techniques and light scattering. They also make use of indirect methods such as thermodynamic properties, conductance and activity studies, diffusion, and viscosity studies. Computer simulation experiments are also used. Since the accurate experimental study of biological solutions is much more difficult than for simple solute solutions, computer simulations may well become a productive study for the complex cases.

AQUEOUS SOLUTIONS OF ELECTROLYTES

Ions in solution are unlikely to be bare: because of ion–solvent interactions $CuSO_4$ in water will form the unit $Cu(H_2O)_n^{2+}$. Hydrated cations and hydrated anions are expected:

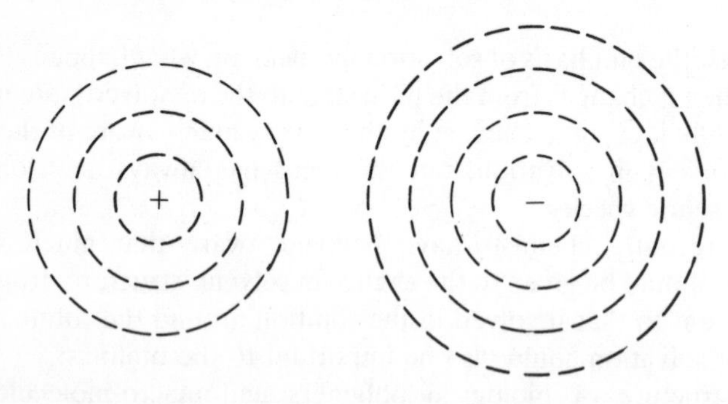

A solvation sphere could be defined as a sphere of water molecules which are closely bound to the ion and move around with the ion as **one single moving entity**, with these solvent molecules having lost their own independent translational motion. The moving ion is thus larger than the bare ion as measured by the crystallographic radius. The bound water molecules contribute to the size of the ion and are very much affected by the ion. Molecules outside the solvation sphere may nonetheless be affected by the ion, with no real sharp cut-off.

We must be very careful to distinguish between:

> regions of the solvent where solvent molecules are actually bound to the ion
>
> and
>
> regions of the solvent which are affected by the ion but are not bound to it
>
> ***This distinction is vitally important.***

The interactions giving rise to solvation are believed to be typically ion–dipole interactions.

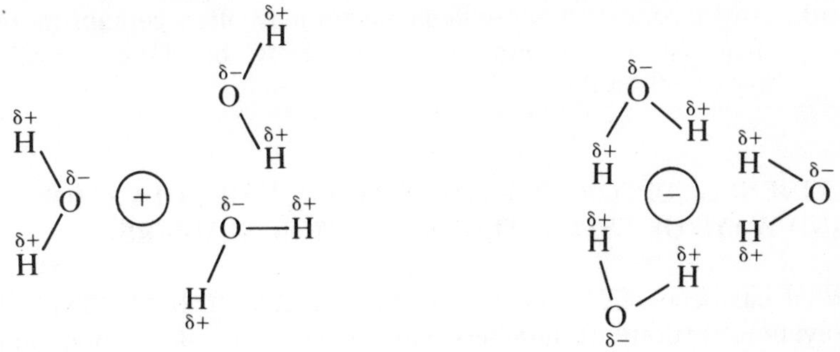

Although we will be concentrating on water as solvent, the same considerations apply to electrolytes in other solvents. Solvation could occur if a sodium salt were dissolved in liquid ammonia; the sodium would be considered to have ammonia molecules closely bound to it giving the unit $Na^+(NH_3)_n$.

Other distinctions which it is crucial to get hold of are those between

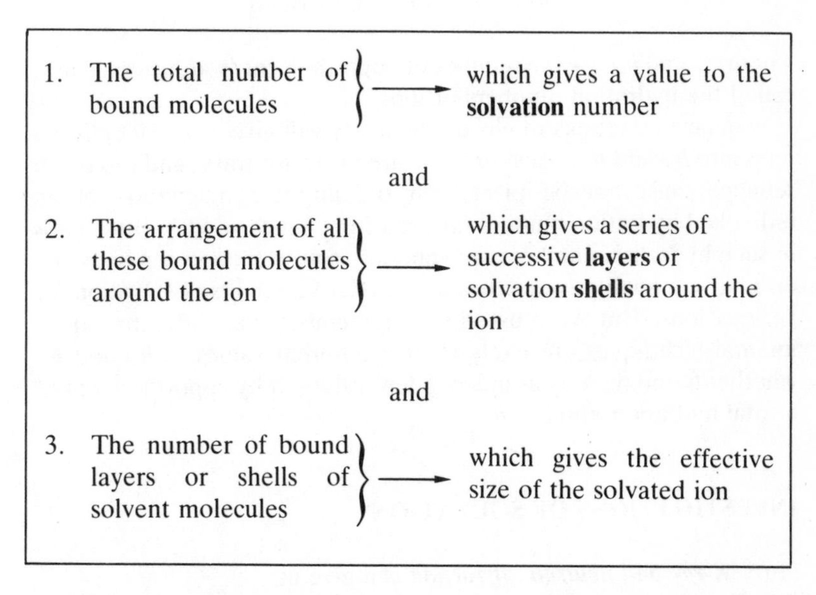

1. The total number of bound molecules
which gives a value to the solvation number
and
2. The arrangement of all these bound molecules around the ion
which gives a series of successive layers or solvation shells around the ion
and
3. The number of bound layers or shells of solvent molecules
which gives the effective size of the solvated ion

The total number of bound solvent molecules are accommodated in these layers around the ion:

(i) For instance, for monatomic ions the first layer often contains four bound solvent molecules if tetrahedrally arranged, six bound solvent molecules if octahedrally arranged.
(ii) Any second and subsequent layers may often contain more than four or six molecules respectively, but little is really known about this.

EVIDENCE SUPPORTING THE EXISTENCE OF SOLVATION AND WAYS OF ESTIMATING SOLVATION NUMBERS

What has been said so far implies that we can attribute individual solvation (hydration) numbers independently to the cation and anion. But not all techniques lead directly to individual solvation numbers, some only give a quantity which applies to the electrolyte as a whole and ways of apportioning this number into a contribution from cation and anion must be devised.

The total solvation number, or hydration number, h, for water is given in the simplest possible manner by:

$$h = h_+ + h_- \text{ for electrolyte MY}$$
$$h = h_+ + 2h_- \text{ for electrolyte } MY_2$$

where h_+ and h_- are the number of bound solvent (water) molecules, called the hydration numbers of ions.

Standard textbooks of electrochemistry will give ways for splitting h up into h_+ and h_-. Most of these are fairly arbitrary, and too much reliance must not be placed on the absolute magnitude of an individual hydration number inferred from a measured h. But trends in such hydration numbers are generally consistent and can be relied on to give a qualitative picture of the relative extents of solvation for various ions. But we must always remember that different experimental techniques are likely to give different values to h_+ and h_- whether found directly as independent values or by apportionment of a total hydration number h.

INVESTIGATIONS OF SOLVATION

(1a) ***X-ray and neutron diffraction*** can give us:

(i) The solvation or hydration number, but so far only for a limited number of cases.

(ii) An estimate of how the solvent molecules are arranged around the ion.
(iii) In a few cases an indication of the number of layers bound to them.
(iv) An indication of the stability of the solvation layers, if synchroton-generated X-rays are used.

Six water molecules are found to be arranged octahedrally around cations such as Ni^{2+}, Na^{+}, and Ca^{2+}, with bound water possibly being confined to one layer only. A similar result has been found for Cl^{-}.
(1b) ***Vibrational spectra (IR and Raman) and electronic spectra (UV, visible and in some cases IR)*** are older techniques but are still of importance. In general the position of an absorption enables us to guess the individual hydration numbers h_{+} and h_{-}. Results suggest that one layer containing six water molecules is fairly typical.
(1c) ***NMR*** is probably the most powerful spectroscopic technique. A shift in the proton resonance frequency and the intensity of the signal can tell us how many water molecules are responsible. Proton relaxation time studies are likely to prove a major advance in the future, and are being progressively applied to solutions containing complex ions such as found in biological solutions. For simple electrolyte solutions the results usually suggest around six water molecules per cation. Individual hydration numbers are obtained.
(1d) ***Ultrasonic relaxation time technique***, described in chapter 2, is a very useful tool which is likely to be further developed in the future.

(2) ***Transport phenomena*** experiments deal with the properties of ions **moving in solution**. Results indicate that the bound water molecules move with the ion, though there are considerable discrepancies in the actual numerical results obtained from each method.

Conductance and transport number experiments study the ion and its solvation shell moving under the influence of an applied field; viscosity and diffusion experiments study the movement of the ion and bound solvent through the solvent.

(a) *Conductance data*

The molar conductances of alkali halides in water vary:

$$\Lambda(\mathrm{LiCl}) < \Lambda(\mathrm{NaCl}) < \Lambda(\mathrm{KCl})$$

which is the reverse of what would be expected if ions were unhydrated. The smallest bare ion, Li^{+}, would be expected to move fastest in an applied field and so be the most highly conducting. This is not so, and it is possible that the moving ions are really $Li^+(H_2O)$ several as against K^+(bare) or $K^+(H_2O)$few. Molar conductances coupled with transport numbers and Stokes' Law give individual hydration numbers for each ion.

Results suggest that if Cs^+ has one layer of water molecules around it, then smaller (on the crystallographic radius basis) ions must have more than one layer around them, for instance one to three layers. Cl^- is believed to have one layer, F^- probably more than one layer, while complex anions such as SO_4^{2-} and NO_3^- have probably less than one layer.

(b) *Viscosity measurements*

A solute alters the viscosity of the solvent. The ion and its water molecules exert a viscous drag on the rest of the solvent and the change in viscosity can be used to calculate a total hydration number which then has to be split up into h_+ and h_-. However, viscosity measurements are normally used as an aid to developing a structural interpretation of solvation.

(c) *Diffusion coefficients*

These again study the ease of movement of the ion through the solvent, and a large variety of techniques can be used. Total hydration numbers are found.

(3) ***Other bulk property measurements***

(a) *Compressibility of the solution*

Compressibility measurements are highly accurate and yield interesting information.

We assume that the bound water molecules are highly compressed by the intense field of the ion and that further compression affects only the unbound water. The decrease in volume accompanying further compression is proportional to the amount of unbound water, and thence can tell us how much bound water there is. This leads to a total hydration number.

(b) *Density measurements*

The density of the pure solvent and the solution can be measured easily and very accurately. Changes in density between pure solvent and solutions can be related to the volume of bound solvent and thence can be used to estimate total hydration numbers.

(c) *Relative permittivity of the solution*

The relative permittivity measures the alignment of the solvent dipoles and production of induced dipoles by an electric field. An ion produces an intense field on bound solvent molecules, and will cause partial if not complete alignment of the dipoles of the solvent molecules affected by the ion. This results in a drop in the observed bulk relative permittivity of the solution relative to the pure solvent. This drop is related to the number of bound solvent molecules. Controversy exists as to whether the effect is restricted to bound molecules only, or whether other solvent molecules are involved.

(4) ***Activity measurements***
This has always been an important, though very indirect, source of hydration numbers. Colligative properties and EMFs give high precision experimental data from which the activity of the solute can be calculated and the stoichiometric mean activity coefficient found. If we assume that solvent molecules are bound to the ions, a relation between the **observed** stoichiometric $f_{\pm}$, the $f_{\pm}$ **calculated** from the Debye–Hückel theory, and the hydration number for the electrolyte as a whole, can be derived. The hydration number h can thus be calculated from a comparison of experiment and theory, and then be split up into h_{+} and h_{-}.

Deviations from predicted behaviour are here interpreted in terms of solvation, but other factors such as ion association may also be involved. Ion association leads to deviations in the opposite direction and so compensating effects of solvation and ion association may come into play. The deviations may also be absorbing inadequacies of the Debye–Hückel model and theory. And so no great reliance can be placed on the actual numerical value of the numbers emerging. This major method is now being largely superseded by the more direct diffraction experiment.

(5) ***Theoretical calculations: computer simulations***
These are being carried out in ever increasing detail, and this technique coupled with diffraction studies may give a means to unravelling the intricacies of solvation.

The technique has been described in chapter 3. In summary, we decide on a model of solvation. The computer uses Monte Carlo or molecular dynamics methods and a simulation of the solvation patterns emerges. The beauty of the method is that we can vary our model at will by varying the type and number of interactions considered for each model. This means we can get a simulation for **each** set of conditions which can be compared with each other and with results from all the experimental methods described earlier. In effect we use the computer to help us to find a model which fits experiment.

CAUTIONARY REMARKS ON THE SIGNIFICANCE OF THE NUMERICAL VALUES OF SOLVATION NUMBERS

Gross discrepancies in the values of h, h_+ and h_- for any given electrolyte are found for the various methods. These may be a result of some or all of the following factors.

(a) Different methods may measure different things: diffraction and spectroscopic methods deal with an equilibrium situation whereas transport properties deal with the moving ion and its bound solvent molecules.
(b) Some experiments could measure bound molecules only, some could measure bound plus affected molecules.
(c) Some of the discrepancies may be absorbing inadequacies of the theoretical model and the equations used in the calculations, for example.
    (i) inadequacies of the Debye–Hückel model and theory in activity methods,
    (ii) inapplicability of Stokes' Law in transport methods,
    (iii) assumptions that the ion and its bound solvent molecules are incompressible in the compressibility studies,
    (iv) lack of knowledge about the varying local relative permittivity in dielectric methods.

SIZES OF IONS

Deciding on the actual size of an ion in solution is one of the fundamental problems of electrolyte solutions. The effective size of an ion appears in many theoretical expressions, but, as yet, there is no theoretical way of calculating it. Nor can it be measured experimentally.

If we can assume an individual solvation number for an ion, and assume an arrangement of water molecules around the ion, we could work out the number of bound layers around the ion. From this the effective size of the ion could be found, provided we assume a volume for each bound water molecule.

However, with all the recent developments it is much better to use both diffraction and computer simulation to work out ion sizes.

A MODEL OF SOLVATION — THE THREE-REGION MODEL FOR AQUEOUS ELECTROLYTE SOLUTIONS

Around the ion is a region of bound water molecules forming layers (or shells) which are highly oriented by the ion–dipole interactions between the ion and the polar water molecules. The arrangement of these bound water molecules is considerably distorted from that of the pure solvent. Next to this region and separating it from the unmodified solvent is an intervening region of water molecules whose structural arrangement changes gradually from the highly oriented bound water to that of unmodified water. This is a region of 'misfit' between bound and unmodified water, and is expected to be highly disordered because of this misfit.

We can compare:

(i) the degree of ordering in ordinary pure solvent

with

(ii) the degree of ordering in the electrolyte solution in terms of three contributions:
   (a) the highly ordered region around the ion,
   (b) the disordered intervening region of misfit,
   (c) the order of the unmodified solvent.

Whether an ion will cause an overall increase in order in the solution compared to the pure solvent depends critically on the balance between the order/disorder introduced by (a) and (b).

If on balance

region (a) is **dominant** there will be an **overall** increase in order in the solution relative to pure solvent, the degree of 'structure' will be enhanced and the ion is termed a '**structure maker**'.

If, on the other hand, on balance

region (b) is **dominant** there will be an **overall** decrease in order in the solution relative to pure solvent, the degree of 'stucture' will be decreased and the ion is termed a '**structure breaker**'.

A cautionary word is necessary here:

> The structure we are talking about is not the structure of ordinary water. The structure which ions enforce on the bound water is a spherically symmetrical arrangement around the ion, and is totally incompatible with the tetrahedral arrangement dictated by the H-bonding array found in pure water. In the intervening 'misfit' region the arrangement is neither a spherically symmetrical nor a tetrahedral one, but will be something which has to change from one extreme to the other, and this could occur over several layers of water molecules. In these layers the molecules must be structurally and dynamically perturbed.

STRUCTURE-MAKING AND STRUCTURE-BREAKING IONS

For a **structure maker** the ordering of the bound water molecules must outweigh the disorder induced in the 'misfit' region. Such ions will be highly polarising, and so are expected to be smaller and highly charged as exemplified by Li^+, Mg^{2+}, F^- and polyvalent cations.

Structure-breaking ions bind water molecules around them, sufficient to cause a 'mismatch' between the structure of the bound water and that of the unmodified water typical of the pure solvent. But the order induced is not great enough to outweigh the disorder created in the 'misfit' region, and this will occur with the less strongly polarising ions such as Rb^+, Cs^+, Br^- and I^-.

Polyatomic ions are more difficult to classify without recourse to experimental clues. But all indications suggest that, possibly because of charge effects

(a) SO_4^{2-} and PO_4^{3-} are powerful structure makers,
(b) NO_3^-, ClO_4^- and SCN^- are structure breakers.

Tetraalkylammonium ions, long chain fatty acid anions, detergents and many biological ions are in a class by themselves. They appear generally to be structure makers, but modern work has shown that their structure-making behaviour is not a result of the molecular processes proposed for simple ions, but is a result of a totally different physical process called hydrophobic hydration.

EVIDENCE FOR STRUCTURE-MAKING/STRUCTURE-BREAKING

This comes mainly from thermodynamic studies.

(a) *Entropies of hydration*

These compare the entropy of solvation of the electrolyte with that for the corresponding number of noble gas atoms or non-polar molecules. For instance, KCl is compared with **two** argon atoms while $CaCl_2$ could be compared with **three** CH_4 molecules.

The processes

noble gas (g) → noble gas (aq)
non-polar molecule (g) → non-polar molecule (aq)

have **large negative** entropies of solvation. This is surprising since it means that the presence of a non-polar inert solute in water causes a considerable **increase in order** and must also alter the structure of the water. It also suggests that **any solute** in water alters its structure. These processes are, therefore, used as a base-line for comparison with solvation of electrolytes.

If the entropy of solvation of an electrolyte has a **more negative** value than that expected, then it is a '**structure maker**'.

If the value is **less negative** than expected it is a '**structure breaker**'.

(b) *Volume changes on solvation*

Analogous interpretations are given for changes in volume on solution of an electrolyte, though the effects are less definite than with the entropies.

If the decrease in volume is **greater** than expected, then the ion is a '**structure maker**' — constriction of the solvent in the bound shell more than outweighs the effect of the 'misfit' region.
If the decrease in volume is **less** than expected, then the ion is a '**structure breaker**' — the constriction of the loosely bound water molecules in the bound region is not sufficiently great to outweigh the effects of the 'misfit' region.

(c) *Viscosity data* (this is a less clear-cut situation)

An ion with a strongly bound inner region will give a **more ordered** solution and hence an increase in viscosity **greater** than expected. This is found for '**structure makers**'.
An ion with a loosely bound inner region will give a **less ordered** solution and hence a **lower** viscosity than expected. This is found for '**structure breakers**'.

(d) *Spectroscopic data*

Raman spectra have been used fairly extensively and NMR can give mean lifetimes for water molecules within the bound layer which are consistent with values obtained from relative permittivity studies and diffusion rates. A wide variety of lifetimes are found ranging from 25 000 ps for a water molecule near Mn^{2+} to 10 ps for Na^{+}, 5 ps for Cl^{-} and 4 ps for I^{-}, and these can be compared with 3 ps for the average lifetime of a water molecule at a given site.

The model described is expected to be an oversimplified description of what happens in solvation. But it has focused attention on to the molecular description and microscopic detail of what happens around an ion in solution, and has thrown into perspective the relevance and importance of diffraction and computer simulation experiments.

This is a book primarily on aqueous electrolytes, but it is imperative that we look at the solvation of non-polar solutes. This is a topic of prime concern for biologists, but it is also very important for chemists since many of the phenomena observed for non-polar and apolar solutes bear a superficial similarity to that for electrolytes. However it is important to point out that despite this similarity the physical and molecular details are totally different for ions and apolar solutes. A study of the reasons for the difference in the molecular behaviour involved in the solvation of ions and apolar molecules has resulted in an increase in our understanding of solvation phenomena in general. It has also led to an upsurge in the number of investigations, both experimental and theoretical, carried out in the last decade.

SOLVATION OF NON-POLAR AND APOLAR MOLECULES — HYDROPHOBIC EFFECTS

When a non-polar or an apolar molecule is dissolved in water we would not expect there to be any interactions of the molecule with the dipoles of the water molecule. Yet, we have seen that solution is accompanied by a **large negative** change in entropy which can only be explained by saying that the solute is doing something to the solvent structure which results in a large overall increase in order.

This observation forces us to ask the question:

What large change can an 'inert' molecule impose on the structural behaviour and properties of pure liquid water?

The only feasible answer is that the normal extensive H-bonded structure of water responds to the presence of the solute molecule by spatial and orientational rearrangements which result in an overall increase of order on formation of the solution. The loose H-bonded structure of ordinary water is lost, and in its place a new H-bonded structure is set up where the water molecules form a cage-like structure around the solute molecules. A large number of cage structures can be formed. These differ in size and shape, and can therefore accommodate different sizes of solute molecules, but all cage structures are made up of H-bonded water molecules. When a non-polar or apolar solute is solvated in this way by having a cage of solvent molecules around it, we talk about **hydrophobic hydration**. Hydrophobic hydration is only possible because liquid water can exist in many different spatial H-bonded arrangements **over and above** the open and extensive tetrahedral H-bonded array of ordinary water. Hydrophobic hydration generally only occurs with **non-polar** and **apolar** solutes, though alcohols, ethers and amines must be included in an exhaustive study. Polar solutes, we have seen, break down the H-bonded arrangement of water and replace it by a spherically symmetrical non-H-bonded shell of water molecules.

Hydrophobic hydration involves relatively **rigid** cages around molecules of solute. In this way, a structure is formed that is more ordered than ordinary water. The observed large negative $\Delta S^{\ominus}$ for hydration is thought to be a consequence of this.

What you must realise is that hydrophobic hydration is **physically** a consequence of rearrangement of the water structure around the solute, and is **NOT** primarily a result of direct attractive interactions between solute and water molecules. It is a hydration phenomenon totally distinct from the sort of attractive interaction hydration which occurs when a polar or charged solute is dissolved in water (often called hydrophilic hydration in biochemical contexts).

EXPERIMENTAL TECHNIQUES FOR STUDYING HYDROPHOBIC HYDRATION

These can be broadly classified into three main categories:

(i) spectroscopic, NMR and ultrasonic relaxation,
(ii) computer simulations,
(iii) thermodynamic studies.

Results of Methods (i) and (ii)

Spectroscopic studies strongly support cage-like structures, and show strong similarities to the solid clathrates which are well characterised as being crystalline cage-structures around an inert molecule. NMR relaxation results show that the rotational diffusion motions of the water molecules of the cage are inhibited, but the motion of the solute is not. This would be expected if water molecules are 'fixed' in the cage, but the solute is still free to move around in the space inside.

Computer studies have shed some light on the behaviour of aqueous non-polar and apolar solutions. They can give information on solute–solvent and solute–solute interactions in such solutions. This is a powerful tool for studying hydrophobic phenomena and is limited in principle only by the accuracy of the assumed model and the quantities derived from this model which are used in the computer simulations. Simulation is of particular importance in the solution chemistry of large macro-molecules and polymers which are extremely difficult to study experimentally, especially in dilute solution. They are thus likely to be a dominant feature in the future study of hydrophobic hydration.

Computer work so far indicates:

(a) Agreement with other techniques which suggest that there is disruption of the normal H-bonded water structure on introduction of an apolar or non-polar solute. This is replaced by a cage-like H-bonded structure with the solute molecule sitting in the cavity at the centre of the cage.
(b) That cages of different sizes can result from different H-bonded arrangements of water molecules.
(c) That the H-bonds around the centre of the cage may be stronger than those in bulk water.
(d) That the rotational motion of the water molecules in the cage is impeded, and this affects viscosity and diffusion rates.
(e) That thermodynamic functions calculated from computer simulations can be fitted to experimental values.

Results of Thermodynamic Studies

These have been the backbone of studies on hydrophobic hydration. Thermodynamic functions such as ΔG, ΔH and ΔS extrapolated to infinite dilution give information about solute–solvent interactions, whereas the same functions studied over a range of concentrations give information on solute–solute interactions (see chapter 1). Thus

Results on thermodynamic functions at infinite dilution give information on hydrophobic hydration.

(a) *Entropies of hydration at infinite dilution, ΔS°, for hydration*

We have already discussed these values and their implications.

ΔS° is always **large** and **negative** for hydrophobic hydration.

(b) $\Delta V^{\ominus}_{\text{solvation}}$ *at infinite dilution*

This is a change in volume at infinite dilution when the solute and solvent (initial state) are mixed to give the solution (final state). All solute volumes used are for liquid solutes, even though the solute may normally be a gas. This is done because if **all** solute volumes are for the solute as a liquid, comparison is easier and more systematic.

When this is done it is found that

> $\Delta V^{\ominus}_{solution}$ (solute as liquid) is always **negative** for hydrophobic hydration.

If the solute is small we could expect it to fit into the spaces of the normal tetrahedral H-bonded structure of ordinary water, giving a decrease in volume. **But** if cage structures result we could expect the water structure to be compacted around the solute, resulting in a further decrease in volume.

(c) *Thermodynamic transfer functions*

When an apolar or non-polar solute is dissolved in a variety of solvents, water is **always anomalous**. So it is useful to compare properties in each solvent with those in water. And so ordinary thermodynamic functions like $\Delta H^{\ominus}_{solution}$ are converted into transfer functions which represent the change, for instance, in enthalpy for the process of transferring the solute from the given solvent to water (in all cases) at infinite dilution.

(i) Most non-polar and apolar molecules are less soluble (where solubilities are given in activities) in water than in other solvents; therefore $\Delta G^{\ominus}_{t}(s \rightarrow w)$ should be **positive** and this is found experimentally to be so.

> The solute is **much happier** to remain in **solvent s** than to be in **water as solvent**.

In $\Delta G^{\ominus}_{t}(s \rightarrow w)$: t means a change in G for the transfer process; $(s \rightarrow w)$ means transfer from solvent s to water.

(ii) $\Delta H^{\ominus}_{t}(s \rightarrow w)$ is often **negative** and fairly large which tells us thermodynamically that

> the solute molecule is now **energetically happier** to be in **water as solvent** than in **solvent s**.

(iii) The only way that

$\Delta G^{\ominus}_{t}(s \rightarrow w)$ can be **positive**

while

$\Delta H^{\ominus}_{t}(s \rightarrow w)$ can be **negative** and large is for

$T\Delta S_t^{\ominus}(s \rightarrow w)$ to **compensate out**

which means that

$T\Delta S_t^{\ominus}(s \rightarrow w)$ has to be **negative** and **greater** than $\Delta H_t^{\ominus}(s \rightarrow w)$.

This is found experimentally and suggests that

the solute must be passing from a **less ordered solution s** to a **more ordered aqueous solution**.

This is very difficult to explain other than in terms of hydrophobic hydration, that is, cage formation of tetrahedrally H-bonded water molecules around the solute.

To summarise

Thermodynamic functions which are used to characterise hydrophobic hydration are

$\Delta G_t^{\ominus}(s \rightarrow w)$ is positive; $\Delta V^{\ominus}$ hydration is negative $\Delta H_t^{\ominus}(s \rightarrow w)$ is often negative; $\Delta S^{\ominus}$ hydration is large and negative $\Delta S_t^{\ominus}(s \rightarrow w)$ is negative;

HYDROPHOBIC HYDRATION FOR LARGE CHARGED IONS

Tetraalkylammonium ions are interesting. At first sight they appear to be structure-making ions but, in fact, their properties are due to hydrophobic hydration as shown by thermodynamic, spectroscopic, conductance and viscosity data. Computer simulation experiments also fit in with the ideas of hydrophobic hydration. The long alkyl chain acts as a non-polar residue which induces the orientational changes in the water structure characteristic of hydrophobic hydration. It is likely that ions which combine a charge with a substantial alkyl or aryl residue, for example, detergents, phospholipids and many biologically active molecules, will show the same pattern of behaviour. Chapter 5 shows these ions must be thought of in a different way from simple small approximately spherically symmetrical ions. They may well form a bridge between the microscopic behaviour of simple electrolytes exhibiting one type of molecular behaviour to large polyelectrolytes and charged biological molecules and macro-molecules exhibiting an entirely different type of molecular behaviour, **even though** their macroscopic properties may sometimes be similar.

It is worth while to emphasise the crucial distinction between

hydration of ionic solutes

and

hydration of non-polar and apolar solutes

Ionic solutes are hydrated as a consequence of attractive electrostatic interactions between the charges on the ions and the dipoles of the water molecules. These interactions break down the tetrahedral arrangements of H-bonded water molecules near the ion to form a spherically symmetrical hydration shell which is **not** H-bonded in the pattern characteristic of the pure solvent.

Non-polar and apolar solutes are hydrated not because of attractive interactions between the solute molecules and the dipoles of the water molecules, but because the presence of the solute induces rearrangements of the H-bonds. Hydration of these solutes is possible only because water can form another type of tetrahedral array – cage arrays. But in these cages the solute molecule is surrounded by water molecules which are **still** H-bonded together and are **tetrahedrally** disposed.

HYDROPHOBIC INTERACTION

When finite concentrations of apolar molecules are considered, it would be expected that there would be no attractive interactions between the apolar solute molecules. However, when the concentration dependence of various thermodynamic functions is looked at (concentration dependence gives information on solute–solute interactions), the behaviour observed is not consistent with that expected for non-interacting solute particles. For instance the **entropy of solution** is found to be

less negative than would be predicted on the basis of the limiting $\Delta S^{\ominus}$ (discussed in hydrophobic hydration) and the increasing number of particles present in solution.

The presence of supposedly non-interacting solute molecules appears to be reducing the large negative entropy change associated with hydrophobic hydration, and is partially reversing the entropically unfavourable hydrophobic hydration.

Observations on other thermodynamic functions collected at finite concentrations support the idea that something is happening to reverse partially the hydrophobic hydration, and this effect manifests itself more obviously as the concentration increases. This effect is called the **hydrophobic effect** or **interaction**. The simplest interpretation pictures two cages each with a solute molecule in its cavity coming close enough to squeeze all the water molecules out from between the cages, and to enable the solute molecules to touch. This results in an **increased disorder** corresponding to an **increase in entropy** which fits in with the observation that the entropy of solution is **less negative** than expected. More recent detailed work shows that this picture is oversimplified.

A lot of the evidence supporting the idea of a hydrophobic effect comes from detailed thermodynamic studies over a range of concentrations.

A similar conclusion of too simple a picture of the hydrophobic interaction is reached when the physical properties of apolar molecule solutions are looked at in more detail.

> One crucial point which you must realise is that the hydrophobic effect, like hydrophobic hydration, is physically a consequence of rearrangement of the water structure around the solute molecules. It is **not** a result of a **direct attractive interaction** between two solute particles. It is very important for biologists to see this distinction because both hydrophobic hydration and hydrophobic interaction play a crucial role in determining the structure of biological molecules *in vivo*, and in determining their biological function. For instance, when two apolar parts of a protein appear to be 'attracted' towards each other it is **not** because of an attraction between these two parts of the protein molecule but because overlap of the cage structures around the apolar residues results in a change in the water structure **between** the apolar residues of the protein.

COMPUTER SIMULATIONS OF THE HYDROPHOBIC EFFECT

Computer simulation experiments suggest that the hydrophobic effect is more complex than previously thought. They fit in with the idea of at least two sorts of 'interaction', one corresponding to solute–solute contact with no solvent between the solute molecules, the other corresponding to a solvent-separated hydrophobic effect where the two solute molecules are still solvent separated.

These pictures of the hydrophobic effect are reminiscent of the concepts of inner and outer sphere ion pairs as discussed in chapter 2.

HYDRATION IN THE STABILISATION OF BIOLOGICAL STRUCTURES AND IN BIOLOGICAL FUNCTION

It has been well known since the early 1960s that hydrophobic effects are of importance in the folding of proteins. Attempts have also been made to describe this folding in more detail by looking at the relative hydrophobic and hydrophilic character of the various aminoacid residues in the protein. Proteins are often made up of alpha helixes, beta sheets and stretches of irregular backbone, and these parts all have different hydration behaviour.

Non-biological polymers such as poly(vinyl alcohol) have been used as model systems for hydrophobic and hydrophilic effects in the folding of proteins. Results suggest that changes in patterns of hydration where, **in turn**, hydrophobic hydration, hydrophilic hydration and hydrophobic interaction are dominant, are reflected by major conformational changes in the molecule. Addition of electrolytes affects these hydration patterns, causing concomitant changes in the folding patterns, and analogies with the effect of electrolytes on protein stability are pertinent.

Biopolymers are considerably hydrated whether the molecule is in a predominantly native, often helical, conformation or in its denatured state which is often a random coil. For many proteins, polypeptides, nucleic acids and other biological macro-molecules, the solvent stabilises certain conformations in preference to the very many other possible conformations. For proteins a very broad and general classification is possible.

(a) Conformations where intereactions between peptide units predominate over interactions between peptide group and solvent — corresponding to the native helical conformation.

(b) Conformations where peptide–solvent interactions predominate — corresponding to the denatured random coil configuration.

One suggestion is that the change from the helical to the random coil conformation involves transfer of non-polar side change peptide groups from the interior of the native helix to the surface of the random coil. In the helix these side chains are forced into the interior by the water and are therefore protected to a large extent from the solvent. But when they are on the surface as in the coil structure they are exposed to the solvent. These groups are essentially hydrocarbon and are basically non-polar residues. When they appear on the surface of the protein they will be involved in hydrophobic hydration. In the interior the converse process of contact between these non-polar side chains results in the elimination of solvent characteristic of hydrophobic interactions. But proteins also contain polar and charged side groups and these will be hydrated in a manner similar to that for simple ions where there is interaction of the group or charge with the dipoles of the water molecules. With any conformation changes there will be resultant changes in the hydration of these polar and charged groups. The final conformation taken up by a protein in any given situation is a balance of many complex processes, but there is now no doubt that it is very much influenced by hydration effects.

Concepts of hydrophobic effects are frequently used in discussions involving membrane and micelle formation. Membranes are made up of lipids and their biological functioning is critically dependent on their hydration behaviour.

Ions forming micelles are hydrophobic as a result of the long hydrocarbon chain. Micelle formation is a balance between the following:

(a) Repulsive interactions between one charged group at the end of a chain and another charged group on another chain.
(b) Attractive interactions between the charged groups on the chains and the counter ions.
(c) Interactions between the solvent and the charged group on the chain and between the solvent and the counter ions — 'ordinary' hydration.
(d) Hydrophobic hydration of the alkyl chains and hydrophobic interactions between the chains.

It is very difficult to get experimental evidence as to the distribution of water **around** and **in** a biological macro-molecule. However, very recent X-ray techniques and analytical procedures are enabling water molecules to be pin-pointed around the molecule. Enzymes and other structures such as genetic material are being studied in this way. Many of these macro-molecules are charged at various positions along the structure and small counter ions will be present to balance the charge on the biological molecule. Carboxylate groups can be present, for instance, as the Ca^{2+}, Mg^{2+} or Na^{+} salts, and protonated amino groups will be balanced by ions such as Cl^{-} or OH^{-}. Water molecules hydrating these groups can be detected and the structure of the water around these ions is similar to that described for simple ions. Location of discrete water molecules at various places along the chain and associated with the non-polar parts suggests involvement of hydrophobic hydration in the short-range features of folding in the chain. Water molecules also provide 'bridges' between the charged groups, the counter ion, and another charged group, or between aminoacid residues through hydrophobic hydration of the apolar parts. These 'bridges' are thought to be involved in the chemical reactions of the macro-molecules, for example by providing pathways for reaction to occur, in particular for proton transfer between various sites. Water around the protein is much more difficult to pick up and is now believed to affect the global stability of the macro-molecule as discussed earlier for proteins.

Haemoglobin has been studied extensively. It has two conformations: one in the presence of O_2 and another in the absence of O_2. The transition between the two conformations can be related to the exposure of hydrophobic parts of the molecule to the solvent. Hence the ability of haemoglobin to transport O_2 may well be crucially related to hydration patterns.

Theoretical studies on the transition between native and active states of proteins and the inactivated denaturated state show that it is very easy to disrupt the active state, and that the equilibrium is very sensitive to hydration interactions and thence to the water structure. All of the types of hydration are particularly sensitive to very slight changes in the environment, for example pH, temperature, pressure, electrolyte concentration and solvent composition. This has been well known experimentally but is now being implicated in the theoretical work.

Structure-breaker ions and molecules are particularly good at destabilising many native states and causing loss of biological function. This must be related to their effect on the water structure which in turn affects the hydration patterns in the macromolecule.

Conversely, molecules or ions which are hydrophobic can stabilise native states, possibly by hydrophobic interactions involving apolar aminoacid residues located on the protein structure.

HYDROPHILIC HYDRATION

This is a hydration phenomenon which occurs when polar molecules which have groups capable of forming hydrogen bonds are dissolved in water. Since many complex molecules which are important biologically contain $-OH$, $-COOH$, $-NH_2$ groups as well as peptide, glycosidic and phosphate linkages, we can expect hydrophilic hydration to be of importance in the solution chemistry of such molecules. Examples are carbohydrates, proteins, nucleotides and even more complex molecules containing carbohydrate, protein and such like residues.

All of these molecules form hydrogen bonds with water molecules through the oxygen or nitrogen atoms contained in the molecule. These hydrogen bonds can result in both large and subtle effects appearing in the aqueous chemistry of solutions found in biological systems. For instance, hydrogen bonding can affect solubility, wetting properties and conformations. Hydrogen bonding can also give rise to quite subtle differences in the behaviour of very similar molecules.

It is hoped that this short discussion will have emphasised to you the crucial importance of all types of hydration phenomena to the understanding of the behaviour of aqueous solutions of biologically important molecules.

5 Charge-separated Ions

So far the emphasis in this book has been on the properties and microscopic behaviour of simple ions such as Mg^{2+}, NO_3^- or even $Cu(NH_3)_4^{2+}$ though at various points we have made reference to charge-separated ions such as

$$(CH_3)_3\overset{+}{N}CH_2CH_2CH_2\overset{+}{N}(CH_3)_3 \text{ or } \overset{+}{N}H_3CH_2C\begin{matrix} /\!\!/ O \\ \backslash O^- \end{matrix}$$

and have hinted that such ions might well behave rather differently from the simple ions like Zn^{2+} or S^{2-}.

We have also mentioned that complex biological macro-molecules often have charges separated by an alkyl chain, or have charges which in terms of the backbone of the molecule seem to be far apart but, because of the particular conformation taken up by the molecule, are in fairly close proximity to each other. Many macro-molecules are polyelectrolytes with charged groups at many points in the macro-molecule. These polyelectrolytes are very complex structures, and are a far cry from the simple inorganic ions studied by the solution chemist.

However, the relatively small charge-separated ions which we will be talking about in this chapter could well form a very useful bridge between the simple ion and the complex polyelectrolyte. It is quite obvious from experimental studies that charge-separated ions are distinct in their behaviour, and that we must modify our ideas and theories on the electrolyte behaviour of simple ions to accommodate the behaviour of charge-separated ions. If we have to do this for basically simple charge-separated ions, it is obvious that for biological polyelectrolytes a major re-think will be necessary. However, charge-separated ions are easier to deal with experimentally and their macroscopic behaviour thereby simpler to explain. The first step towards building up an explanation and description of the molecular behaviour of biological polyelectrolytes could well be based on charge-separated ions. Unfortunately, not a great deal of work has been done experimentally, and virtually nothing has been done on the adaptation of our existing theories for simple electrolyte solutions.

A look at the sort of polyelectrolytes and macro-molecules which biochemists and biologists study could help to put the whole matter into context.

TYPICAL BIOLOGICAL MOLECULES

There are three major groups of biological polymers — the carbohydrates, the proteins and the nucleic acids together with groups of copolymers such as glycoproteins and nucleoproteins. Between them these cover a vast range of biological material.

Carbohydrates are polymers of simple sugars and have large numbers of —OH groups some of which are replaced by sulphate, carboxylate or phosphate groups such as:

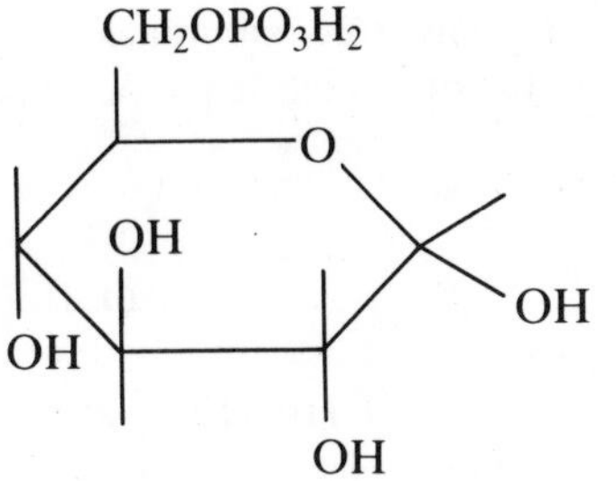

$CH_2OPO_3H_2$

O

OH

OH

OH

OH

here the phosphate can be ionised to give a negative charge

glucose 6-phosphoric acid

and some —OH groups can be replaced by —NH_2 groups as in

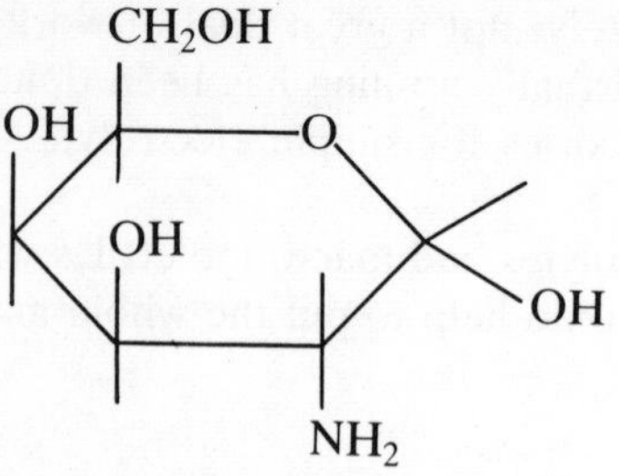

CH_2OH, OH, O, OH, OH, NH_2

galactosamine

here the amino group can be protonated to give the $—\overset{+}{N}H_3$ group

Because of substitutions like these, a polysaccharide could have charges along its backbone.

Proteins are polymers of amino acids and have a

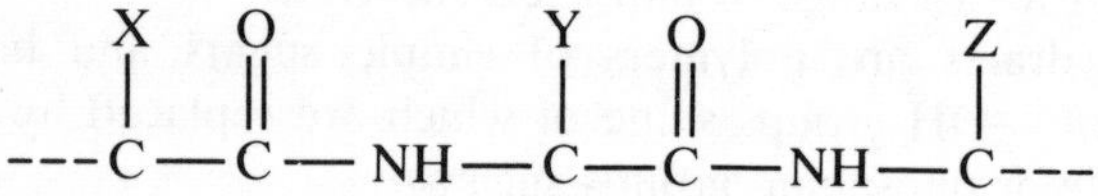

$$---\overset{X}{\underset{|}{C}}—\overset{O}{\underset{||}{C}}—NH—\overset{Y}{\underset{|}{C}}—\overset{O}{\underset{||}{C}}—NH—\overset{Z}{\underset{|}{C}}---$$

backbone where X, Y, Z and so forth represent the side chains of the aminoacids making up the protein. The free aminoacids contain —NH_2 and —COOH groups, both of which can be charged. Examples are

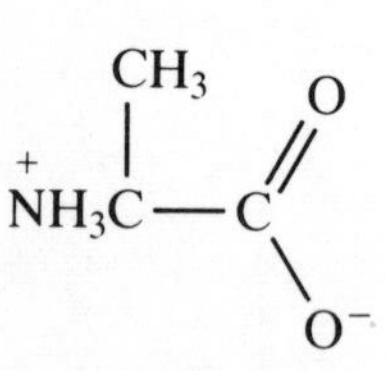

CH_3, $\overset{+}{N}H_3C—C(=O)O^-$

alanine

$CH_2CH_2C(=O)NH_2$, $\overset{+}{N}H_3C—C(=O)O^-$

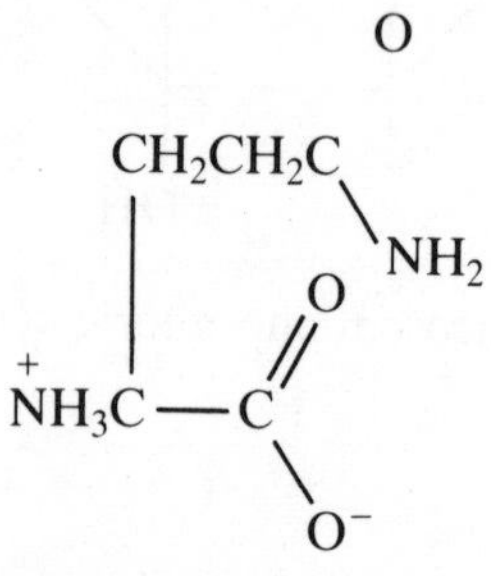

glutamine

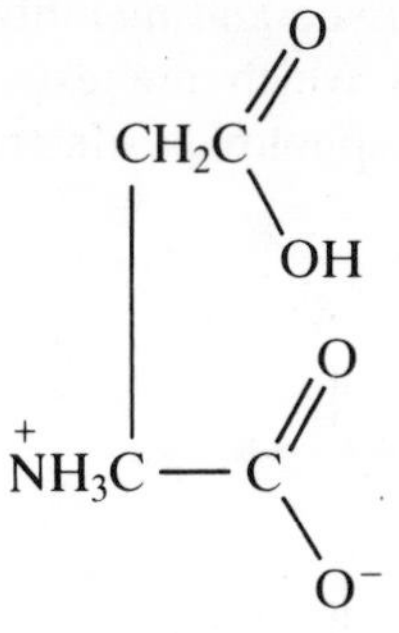

aspartic acid

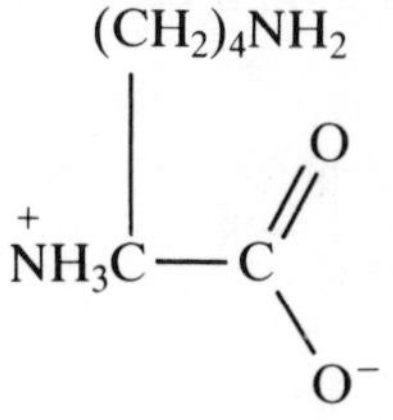

lysine

So, like the carbohydrates, proteins can have charges along the molecule, though these occur only on the side chains.

Nucleic acids are polymers of sugar molecules bridged by phosphate groups where the sugar molecule has a heterocyclic base attached to it. Charges are possible wherever there is a free phosphate group giving a negative charge and whenever a nitrogen in the heterocyclic base is protonated giving a positive charge. For instance, the nucleotide monomer unit adenosine monophosphate has two ionisable groups which could give a charge-separated ion.

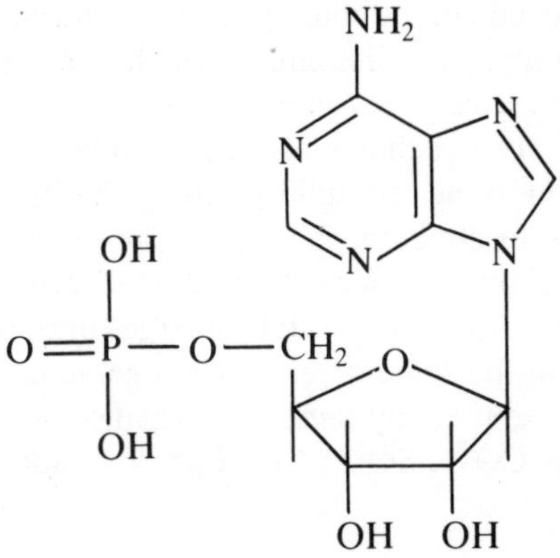

In these biological polymers the separation of the charges will depend on the conformation taken up by the molecule in solution, and this will depend on many factors including electrostatic interactions, hydrogen bonding, hydrophilic and hydrophobic hydration.

Lipids are macromolecules found in fatty tissue and membranes. Phospholipids are similar to ordinary lipids which are esters of glycerol and fatty acids. Two common phospholipids illustrating charge separation are

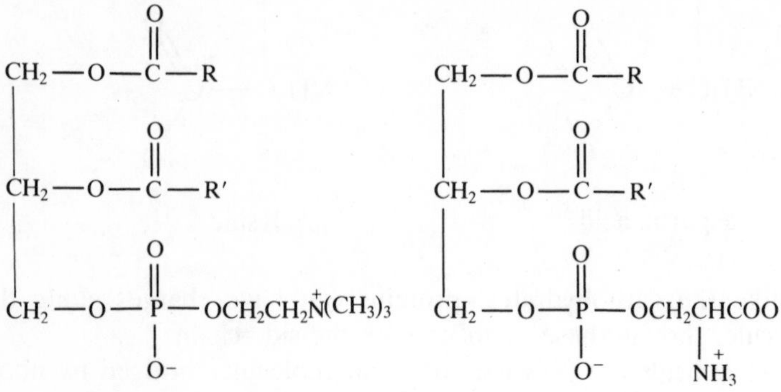

$$
\begin{array}{l}
CH_2-O-\overset{\overset{\displaystyle O}{\|}}{C}-R \\
\;| \\
CH_2-O-\overset{\overset{\displaystyle O}{\|}}{C}-R' \\
\;| \\
CH_2-O-\overset{\overset{\displaystyle O}{\|}}{\underset{\underset{\displaystyle O^-}{|}}{P}}-OCH_2CH_2\overset{+}{N}(CH_3)_3
\end{array}
\qquad
\begin{array}{l}
CH_2-O-\overset{\overset{\displaystyle O}{\|}}{C}-R \\
\;| \\
CH_2-O-\overset{\overset{\displaystyle O}{\|}}{C}-R' \\
\;| \\
CH_2-O-\overset{\overset{\displaystyle O}{\|}}{\underset{\underset{\displaystyle O^-}{|}}{P}}-OCH_2\underset{\underset{\displaystyle \overset{+}{N}H_3}{|}}{C}HCOO^-
\end{array}
$$

phosphatidyl choline (lecithin)   phosphatidylserine

Other examples of potentially charge-separated biologically important molecules are enzymes, the biological catalysts, which are proteins and therefore are polyelectrolytes, and neurotransmitters which pass messages to and from the brain.

Drugs, medicinal compounds, pesticides and herbicides are some of the varied range of compounds which can also act as charged molecules in biological environments.

Interactions between charges are going to be of overwhelming and often crucial importance in understanding the biological function of these molecules in aqueous solutions. The charges on these molecules, or along the surfaces of proteins, genetic materials and membranes must be balanced out by charges of opposite signs. These may be found on the same or on other organic molecules, but there will also be present many simple inorganic ions such as HPO_4^{2-}, $H_2PO_4^-$, HCO_3^-, CO_3^{2-}, SO_4^{2-}, Cl^-, Ca^{2+}, K^+ and Na^+.

A BRIEF DISCUSSION OF THE PHYSICAL CHEMISTRY OF SIMPLE CHARGE-SEPARATED IONS

For simple ions the charge is unambiguous, for charge-separated ions it is not. For ions such as $(CH_3)_3\overset{+}{N}CH_2CH_2COO^-$ the overall charge is zero but the molecule certainly does not always behave as though it were neutral. Compounds such as $\overset{+}{N}H_3CH_2COO^-$ could well function as dipoles, especially where the opposite charges are close together. A compound such as

$$SO_3^- \text{—} C_6H_4 \text{—} SO_3^-$$

looks like a divalent anion, while $(CH_3)_3\overset{+}{N}CH_2CH_2\overset{+}{N}(CH_3)$ looks like a divalent cation, but they may well not act as such.

As the distance between the charges becomes greater, one question becomes particularly relevant. Are there conditions under which the charge will start to simulate separate charges, and each charge start to build up its own ionic atmosphere so that the unit then behaves like two independent charges carried on the same framework?

The radius of an ionic atmosphere gives an indication of how far out from the central ion the most important part of the atmosphere lies. A small radius means a compact ionic atmosphere and a large radius means a diffuse ionic atmosphere. The magnitude of the radius of the ionic atmosphere depends critically on the ionic strength.

Ionic strength (mole litre $^{-1}$)	0.05	0.10	0.50	1.00
radius of atmosphere (Å)	14	9.5	4.0	3.0

At high concentrations the ionic atmosphere radius will become progressively less than the distance of charge separation and it seems reasonable to suppose that when it reduces to around the charge-separation distance the model of a single ionic atmosphere over the whole ion must fail.

Can we, therefore, feel justified in using Debye–Hückel ideas and equations for species such as $SO_3^-(CH_2)_nSO_3^-$, and what do we use for a compound such as $SO_3^-(CH_2)_n\overset{+}{N}(CH_3)_3$?

These matters are relevant to complex biological molecules where there are charges along the backbone, or where specific conformations can bring into close proximity charges which are well separated in terms of the backbone of the molecule. Because the substance is a polyelectrolyte, there will often be high concentrations of simple inorganic ions present to balance the charges on the polyelectrolyte. Furthermore, not all biological reactions occur in dilute solution, some occur in very concentrated solution or even gels. And so ionic strengths can easily be sufficiently high for the radius of the ionic atmosphere to be low enough for each charge to function independently.

RESULTS ON SIMPLE CHARGE-SEPARATED IONS

There is little experimental evidence as to whether charge-separated ions **do** simulate separate ions. Activity coefficient studies, analogous kinetic salt effects and ion association in charge-separated ions are the main source of information.

Activity coefficient studies indicate that ions such as

$SO_3^-(CH_2)_nSO_3^-$ and

SO_3^-

SO_3^-

act as divalent ions in dilute solution, but simulate two separate single charged ions in more concentrated solutions. However, other explanations are possible, and it is difficult to get an unambiguous decision.

This has implications for solvation — do we think about solvation of the compound as a whole entity or do we think about the two charged parts separately, and, if so, what is the solvation pattern for the organic part in between?

Is the picture

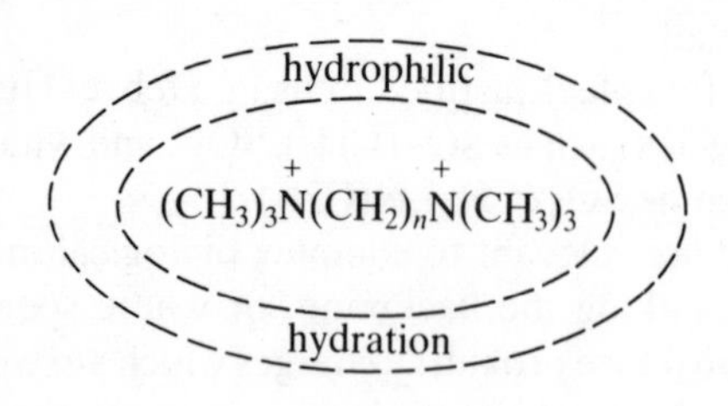

or

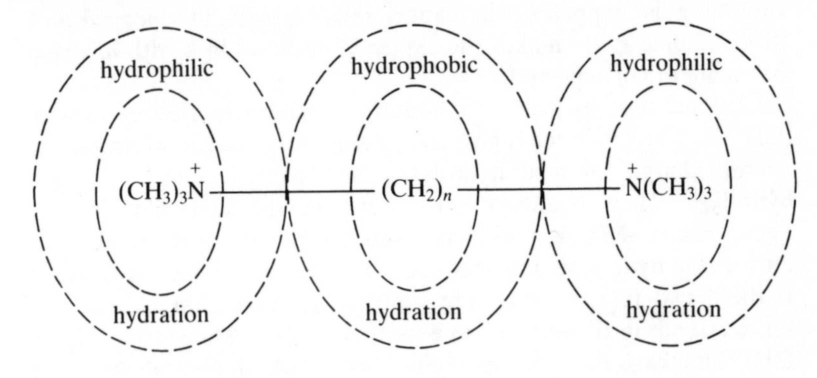

Primary salt effects in kinetics describe the effect of ionic strength on the rate constant for reactions between ions, and are a manifestation of non-ideality of the type dealt with in Debye–Hückel theory. The kinetic analogue of the Debye–Hückel equation for $\log f_{\pm}$ is

$$\log k = \log k_0 + \frac{2Az_Az_B\sqrt{I}}{1 + B\mathring{a}\sqrt{I}} + B'I$$

where k is the observed rate constant at the stoichiometric ionic strength

- k_0 is the extrapolated rate constant at zero ionic strength
- z_A and z_B are the charge and sign on the reacting ions A and B
- $B\mathring{a}$ is as in Debye–Hückel theory
- B' is an empirical parameter.

If z_Az_B is **positive**, like-charges reacting, rate constants increasing with ionic strength are expected.
If z_Az_B is **negative**, unlike-charges reacting, rate constants decreasing with ionic strength are expected.

Values of B' give an indication of any **specific** salt effects. Experiments where one reactant is charge-separated show that these effects (values of B') **increase** with increasing charge separation, and are negative for reactions between like-charges and positive for reactions between unlike-charges.

If z_Az_B is **positive** (like-charges) and B' is **negative**, then something is happening to cancel out partially the normal salt effect expected from Debye–Hückel theory; B' makes the expected increase in k with increase in I smaller than expected.

If $z_A z_B$ is **negative** (unlike-charges) and B' is **positive**, then something is happening to cancel out partially the normal salt effect — that is, B' makes the expected decrease in k with increase in I smaller than expected.

Acid and base hydrolysis of quaternary ammonium esters such as $(CH_3)_3\overset{+}{N}(CH_2)_nCOOCH_3$ give charge-separated transition states: +2 overall charge for acid hydrolysis, zero overall charge for base hydrolysis. There is always a +1 charge on the quaternary ammonium group $—\overset{+}{N}(CH_3)_3$ with the ester site being positively charged during acid hydrolysis and negatively charged during base hydrolysis. If these two regions of charge simulate separate charges, the observed trends in B' would be as expected. In qualitative terms, H^+ or OH^- attacking the ester grouping 'sees' only a neutral molecule instead of a positively charged ion, and this would give the observed behaviour.

For reactions between ions it is the transition state of the reaction which is charge-separated. At higher ionic strengths the radius of the ionic atmosphere becomes comparable with the distance of separation of the charges. The partial reversal of the expected primary salt effects could be a manifestation of the overall charge on the transition state simulating individual ions each with its own ionic atmosphere.

Another aspect of charge separation in the transition state is shown in the base hydrolyses of

$$CH_3COOEt,\ \overset{+}{N}(Et)_3CH_2COOEt,\ (CH_3)_3\overset{+}{N}CH_2CH_2OCOCH_3$$

and the reaction

$$(CH_3)_3CHOHCH_2COCH_3 \rightarrow 2CH_3COCH_3$$

carried out in the presence of Ca^{2+}, Ba^{2+} or Tl^+ where formation of the ion-pairs $\overset{+}{Ca}OH$, $\overset{+}{Ba}OH$ or TlOH will remove substantial amounts of OH^- from solution. A decrease in rate would be expected for reactions carried out in the presence of these ions. However, if $\overset{+}{Ca}OH$ ($\overset{+}{Ba}OH$ or TlOH) were as reactive as OH^-, or, put equivalently, if Ca^{2+} (Ba^{2+} or Tl^+) associated with the transition state formed from the substrate and OH^-, then no change in rate would be found. Experiments looking at these effects can lead to some interesting conclusions regarding the charge distribution in the respective transition states.

The results can be summarised as follows:

Possible charge distribution and structure of the transition state	Overall charge on the transition state	Effect of Ca^{2+}, Ba^{2+} or Tl^+ on the rate
1. 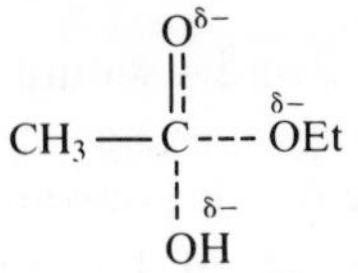 CH_3—C(=$O^{\delta-}$)(---$OH^{\delta-}$)---$OEt^{\delta-}$	−1	No effect, $\overset{+}{Ca}OH$ is as effective as OH^-. Ca^{2+} associates with the transition state
2. 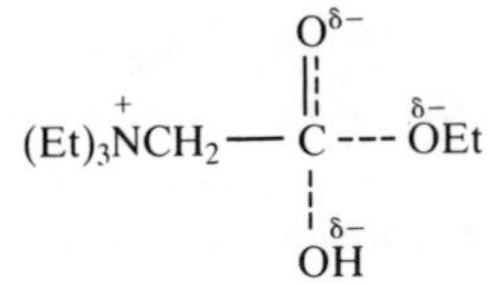 $(Et)_3\overset{+}{N}CH_2$—C(=$O^{\delta-}$)(---$OH^{\delta-}$)---$OEt^{\delta-}$	zero overall	Decrease in rate found. CaOH is unreactive. Ca^{2+} does not associate with the transition state
3. $(CH_3)_3\overset{+}{N}CH_2CH_2$—$O^{\delta-}$---C(=$O^{\delta-}$)(---$OH^{\delta-}$)—$CH_3$ 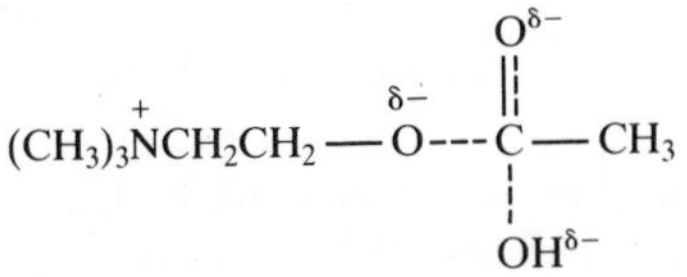	zero overall	No effect. Ca^+OH is as effective as OH^-. Ca^{2+} associates with the transition state
4. 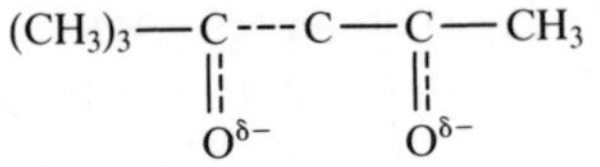 $(CH_3)_3$—C(=$O^{\delta-}$)---C—C(=$O^{\delta-}$)—CH_3	−1	Decrease in rate found. $\overset{+}{Ca}OH$ is unreactive. Ca^{2+} does not associate with the transition state

We have to explain why Ca^{2+} appears in the transition states 1 and 3 but not in 2 and 4.

In 1 the negative charge is concentrated around the reaction site and association with cations may well resemble that of the OH^- ion with cations as in $\overset{+}{Ca}OH$.

In 4 the negative charge is now shared between two distant oxygen atoms, there is considerable charge separation and the Ca^{2+} may not 'see' this as an overall negative charge as in 1.

In 2 and 3 the transition states are overall zero-charged and would not be expected to be attractive to Ca^{2+}. Yet in 3 Ca^{2+} appears in the transition state while in 2 it does not.

The difference in behaviour can be explained in terms of the distance of the positive charge from the reaction site. In 2 the Ca^{2+}

'sees' the positive charge and reaction site negative charge as a total — that is, it 'sees' the overall zero charge — and so is not attracted to the reaction site. In 3 the positive charge is sufficiently far away for it not to be 'seen' by the Ca^{2+} as it 'looks at' the reaction site. The Ca^{2+} 'sees' only the negative charge on the reaction site and associates with it.

It is interesting that the behaviour of Ca^{2+} in 2 and 4 should be so similar, 2 having an overall zero charge **but** being charge-separated. As far as Ca^{2+} is concerned, the transition state of 4 is equivalent to an overall zero-charged transition state such as in the slightly charge-separated 2.

These are kinetic examples of how the overall charge may be relatively unimportant while the charge distribution is of crucial importance.

Recent work has shown that the same considerations apply in ion-association studies where one or both ions are charge-separated.

The interaction of various charge-separated dicarboxylate anions has been studied with simple inorganic cations and with charge-separated organic cations.

It is quite clear from this work that there is a high degree of specificity and selectivity in these interactions. Where spatial charge-matching occurs, enhanced association occurs, and this is apparent even in the divalent anions where the two carboxylate groups are separated by a flexible methylene chain in the series $^{-}OOC(CH_2)_nCOO^{-}$ where $n = 0, 1, 2, 3, 4$ and 6. When the framework between the charges is flexible there is much more latitude in the distance of separation of the two charges, and conformational changes can occur which enable the anion to wrap around the cation so that spatial charge matching can occur. When the framework is rigid and the charge-separation distance is much more nearly fixed, the charge-matching is more specific and more enhanced when favourable charge separation in both anion and cation occur. The rigid dicarboxylate anions studied ran in the series:

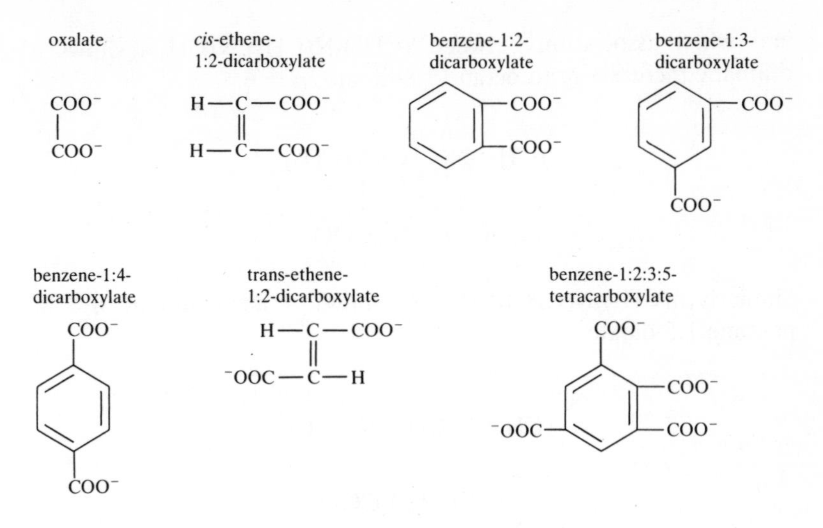

The charge-separated cations were also flexible in

$$(CH_3)_3\overset{+}{N}(CH_2)_n\overset{+}{N}(CH_3)_3 \qquad n = 2, 3, \text{ and } 6$$

and rigid in

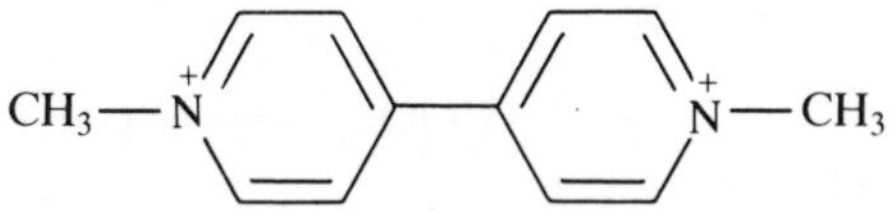

paraquat

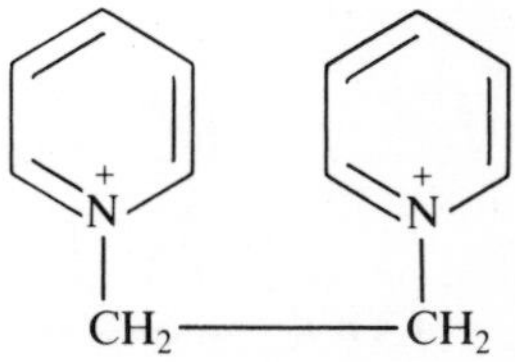

diquat

Fortuitous fitting together of charges and consequent enhanced association is found between diquat and pentane-1:5-dioate, $n = 3$, in the series $^-OOC(CH_2)_nCOO^-$ when a maximum is found at pentane-1:5-dioate

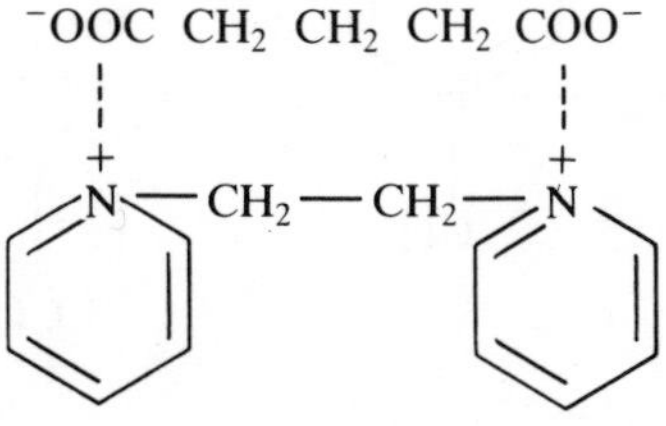

$$^-OOC\ CH_2\ CH_2\ CH_2\ COO^-$$

and when association between $(CH_3)_3\overset{+}{N}(CH_2)_n\overset{+}{N}(CH_3)_3$ shows a dramatic increase up to octane-1:8-dioate, $n = 6$.

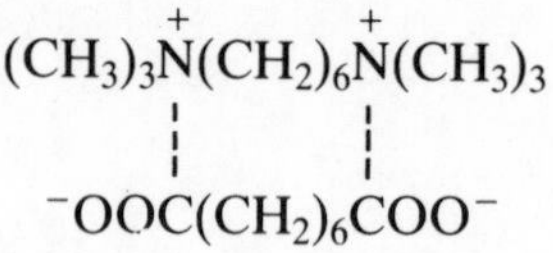

$$\begin{array}{c}(CH_3)_3\overset{+}{N}(CH_2)_6\overset{+}{N}(CH_3)_3 \\ ^{-}OOC(CH_2)_6COO^{-}\end{array}$$

Similarly, $(CH_3)_3\overset{+}{N}(CH_2)_3\overset{+}{N}(CH_3)_3$ shows a maximum at $n = 3$, pentane-1:5-dioate.

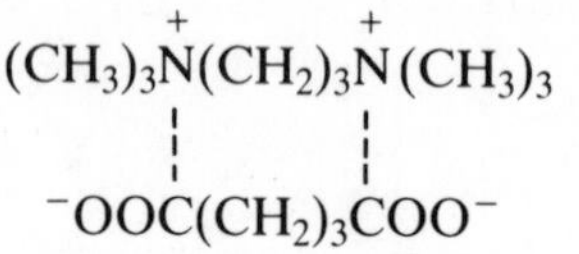

$$\begin{array}{c}(CH_3)_3\overset{+}{N}(CH_2)_3\overset{+}{N}(CH_3)_3 \\ ^{-}OOC(CH_2)_3COO^{-}\end{array}$$

The effects are more obvious with the rigid dicarboxylates; for example, paraquat and $(CH_3)_3\overset{+}{N}(CH_2)_6\overset{+}{N}(CH_3)_3$ show an increase in the series from oxalate to benzene-1:4-dicarboxylate, whereas diquat and $(CH_3)_3\overset{+}{N}(CH_2)_3\overset{+}{N}(CH_3)_3$, which have a smaller charge-separation distance, show no significant change.

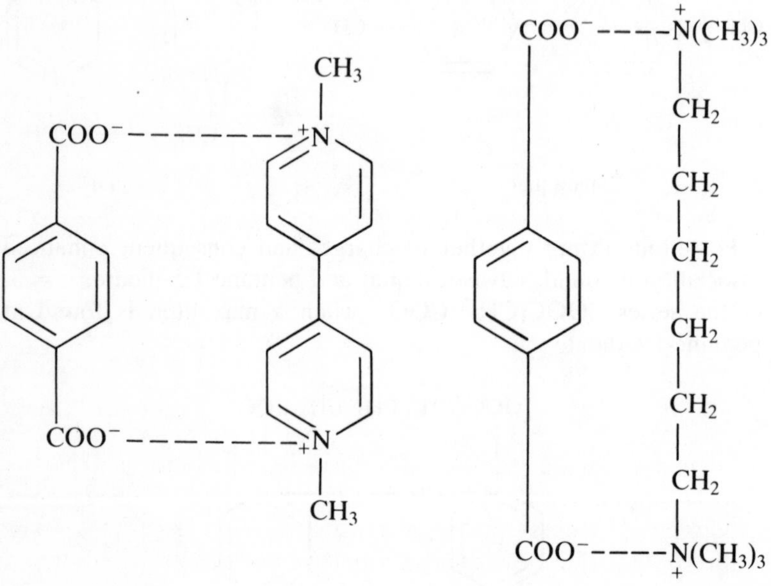

When simple inorganic cations, Mg^{2+}, Ca^{2+}, $[Co(NH_3)_6]^{3+}$ and $[Co(en)_3]^{3+}$ were studied, similar effects were noticed.

For the rigid-framework dicarboxylates a maximum occurs at oxalate for the small Mg^{2+}, but at *cis*-ethane-1:2-dicarboxylate and benzene-1:3-dicarboxylate for the bigger $[Co(NH_3)_6]^{3+}$ and $[Co(en)_3]^{3+}$ respectively.

It is quite clear that again a size-matching effect is operative, and cations which best fit into the spaces between the charges are bound the best. This is particularly well demonstrated in the large association of $[Co(en)_3]^{3+}$ with benzene-1:2:3:5-tetracarboxylate where the $[Co(en)_3]^{3+}$ fits beautifully on top of the large tetracarboxylate, whereas the same cation with oxalate does not give such a good fit and has a much lower association constant.

With flexible anions the same trends are apparent, but are not so dramatic simply because the two carboxylate groups are not held to a rigid distance apart in any one anion and so can accommodate, to a certain extent, to the size of the cation.

A similar effect has been noticed in kinetic studies. In the acid hydrolysis of acetylcholine, disulphonate anions such as $SO_3^-(CH_2)_nSO_3^-$ with $n = 1$ and 2, benzene disulphonates and some naphthalene disulphonates generally show no specific effects, but where 'good' fitting together of the transition state charges and the negative charges in the sulphonate compound is possible, catalysis is observed, as in naphthalene-2:7-disulphonate and naphthalene-1:3:6-trisulphonate.

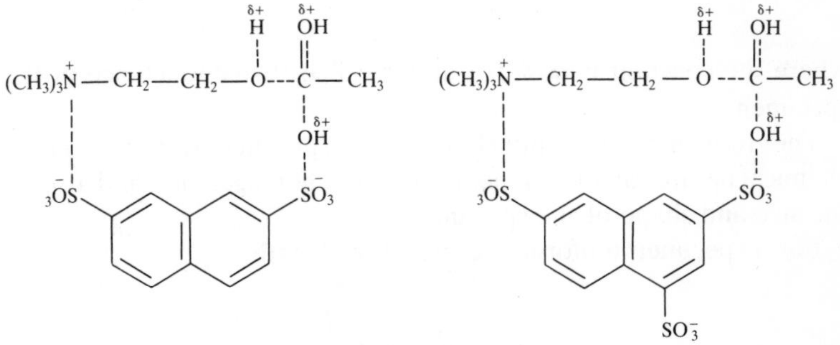

$(CH_3)_3\overset{+}{N}-CH_2-CH_2-\overset{\delta+}{O}H---\overset{\delta+}{C}(OH)(\overset{\delta+}{O}H)-CH_3$ … ^-O_3S–naphthalene–SO_3^- (2:7)

$(CH_3)_3\overset{+}{N}-CH_2-CH_2-\overset{\delta+}{O}H---\overset{\delta+}{C}(OH)(\overset{\delta+}{O}H)-CH_3$ … ^-O_3S–naphthalene–SO_3^-, SO_3^- (1:3:6)

These findings really should not surprise us as they are reminiscent of the geometrical 'fitting together' type of mechanism postulated in many biological reactions. What is perhaps more surprising is that physical chemists have not been more aware of these possibilities for this specific type of ion-association.

Appendix A: Aspects of Physics related to Conductance Studies

Ohm's law can be written as

$$\frac{E}{I} = \text{constant}$$

where I is the current and E is the potential difference across a conductor; the constant is called the **resistance**, R, of the specimen.

$$\frac{E}{I} = R$$

Ohm's Law can also be written as

$$\frac{I}{E} = \text{constant}$$

where the constant is now $1/R$ and is called the conductance of the specimen.

The resistance or conductance of a specimen depends on its chemical nature, on the temperature, on its homogeneity, and also on the size and shape of the specimen.

For a specimen uniform over its whole length

$$\text{resistance } R \propto \frac{l}{A}$$

where l is the length of the specimen
and A is the area of the cross-section.

$$R = \rho \frac{l}{A}$$

where ρ is a constant of proportionality called the **resistivity**.

The same argument can be expressed in terms of the conductance, R^{-1}.

$$\frac{1}{R} \propto \frac{A}{l}$$

and

$$\frac{1}{R} = \kappa \frac{A}{l}$$

where κ is a constant of proportionality called the **conductivity**. Hence

$$\rho = \frac{1}{\kappa} \quad \text{or} \quad \kappa = \frac{1}{\rho}$$

When we deal with the effect of the passage of an electric current through a solution we normally carry through the discussion in terms of the conductivity κ, with

$$\kappa = \frac{1}{R}\left(\frac{l}{A}\right)$$

The experimental measurement which determines κ is the resistance of a given volume of solution.

It is found that κ varies considerably with concentration, and a further quantity is defined, Λ, called the **molar conductance**:

$$\kappa = \Lambda c_{\text{stoich}}$$

where the concentration used is always the **stoichiometric** concentration.

$$\Lambda = \frac{\kappa}{c_{\text{stoich}}}$$

The current is carried in solution by the ions present in the solution, and the greater the number of ions present per unit volume (that is, the concentration) the greater is the current able to be carried, and the smaller will be the resistance. Since

$$\kappa = \frac{1}{R}\left(\frac{l}{A}\right)$$

we would expect κ to increase with concentration and we would expect κ to be roughly proportional to the concentration of ions present in solution.

Since

$$\Lambda = \frac{\kappa}{c_{\text{stoich}}}$$

the molar conductance Λ is the quantity which we can use to enable us to decide whether the **actual** concentration of ions in solution is equal to, or not equal to, that which we would expect on the basis of the stoichiometric concentration.

If an electrolyte solution is considered to be **ideal**, then the current carried would be strictly proportional to the number of ions per unit volume — that is, to the concentration. Hence for ideal solutions we would expect κ to be **strictly** proportional to the concentration of ions **actually** present in solution.

Whether or not the **actual** concentrations of ions present is equal to the **stoichiometric** concentration of electrolyte depends critically on the **type** of electrolyte under consideration.

If the electrolyte is a **strong** electrolyte, it is fully ionised in solution and

$$\left.\begin{array}{l}\text{the actual concentration}\\ \text{of ions}\end{array}\right\} = \left\{\begin{array}{l}\text{the stoichiometric}\\ \text{concentration of ions}\end{array}\right.$$

If the electrolyte is a **weak** electrolyte it is not fully ionised in solution and

$$\left.\begin{array}{l}\text{the actual concentration}\\ \text{of ions}\end{array}\right\} \begin{array}{c}\text{IS NOT}\\ \text{EQUAL}\end{array} \left\{\begin{array}{l}\text{to the stoichiometric}\\ \text{concentrations of ions}\end{array}\right.$$

Relevance of these Arguments to Molar Conductance

$$\kappa = \Lambda c_{\text{stoich}} \qquad \text{(by definition)}$$

therefore

$$\Lambda = \frac{\kappa}{c_{\text{stoich}}}$$

Provided the solution is **ideal**, and this in practice means provided the solution is dilute enough to be considered ideal:

$$\kappa \propto c_{\text{actual}}$$

where c_{actual} is the **actual** concentration of ions, therefore

$$\Lambda \propto \frac{c_{\text{actual}}}{c_{\text{stoich}}}$$

For a strong electrolyte

$$c_{\text{actual}} = c_{\text{stoich}}$$

therefore

$$\Lambda \propto \frac{c_{\text{stoich}}}{c_{\text{stoich}}}$$

hence

$$\Lambda \text{ is a constant}$$

Therefore for a strong electrolyte in ideal conditions we would expect Λ to be independent of concentration.

For a weak electrolyte

$$c_{\text{actual}} \neq c_{\text{stoich}}$$

except under the limiting conditions of **very, very** dilute solutions where the fraction ionised approaches unity.

The fraction ionised α is given by

$$\alpha = \frac{c_{\text{actual}}}{c_{\text{stoich}}}$$

Therefore

$$c_{\text{actual}} = \alpha c_{\text{stoich}}$$

We have seen that

$$\Lambda \propto \frac{c_{\text{actual}}}{c_{\text{stoich}}}$$

that is

$$\Lambda \propto \frac{\alpha c_{\text{stoich}}}{c_{\text{stoich}}}$$

$$\Lambda \propto \alpha$$

Now α is **NOT** a constant over a range of concentrations for a weak electrolyte. In particular, the fraction ionised increases dramatically as the concentration decreases and tends to zero. Hence for a weak electrolyte we would expect Λ to vary with the stoichiometric concentration, and in particular to increase dramatically with decrease in concentration.

Appendix B: Ions — Aspects of Physics necessary for an Understanding of Electrolyte Solutions

Because an ion possesses a charge, the ion can have an effect — generally on substances near it. Near the ion the effect of the charge is greatest, and we talk about the **electric field** of the charge as representing the **effect of the charge**.

We can consider two situations:

(a) an ion taken as a point charge in a vacuum,
(b) an ion taken as a point charge in a medium.

An ion in a **vacuum** produces an effect at a distance, r, from it which depends on the charge and the distance. But if we put an ion into a **material medium** we would find (if we could do the experiment) that the field at any distance, r, from the charge will be **decreased** by a factor whose magnitude depends on the medium.

> But talking in terms of one charge only is a discussion which cannot be related to any actual measurements. If macroscopic measurable quantities are to be related to our conceptual ideas at a microscopic level, a ***second charge must be introduced*** so that we talk about the ***effect of the first charge on the second, and vice versa***.

A second point charge placed at some distance, r, from the first charge results in an **interaction** between them whose magnitude will depend on

(i) the charges,
(ii) the distance between the charges,
(iii) the field at that particular distance,
(iv) whether the charges are in a vacuum or in a material medium: gas, liquid or solid.

CHARGES IN A VACUUM

Consider a charge $+ze$ in a vacuum and place a second charge $-ze$ at a distance r from it. Both charges will have an effect on each other, and at any given point from both charges the **field** will be the resultant sum of the two fields. Furthermore, the effect which each charge has on the other will manifest itself in a **force** between them.

This will give rise to an **energy of interaction** between the two charges, and this electrostatic potential energy of interaction between the two charges is the **work** done in bringing the second charge from infinity to the distance r, between the two charges. The work done is equal to the force between the charges multiplied by the distance through which the force moves. The term **potential** is another way of looking at the effect which one ion can have on another since the potential energy of interaction between two charges can also be thought of as the product of the charge on one ion and the potential at that ion due to the second ion. The potential at a given point can thus be defined in terms of potential energy of interaction between the two ions.

Formally, then:

the electric field, E, due to the **first** charge at the position of the **second** charge $= +\dfrac{ze}{4\pi\epsilon_0 r^2}$

the force, F, between the two charges $= -\dfrac{z^2e^2}{4\pi\epsilon_0 r^2}$

(inverse square law)

the electrostatic potential energy of interaction, u, between the two charges $= -\dfrac{z^2e^2}{4\pi\epsilon_0 r}$

the electrostatic potential energy can also be defined in terms of the potential, ψ

the electrostatic potential energy of interaction, u, between the two charges $= +ze\psi'$ or $-ze\psi''$

where ψ' is the potential **at** charge $+ze$ **due to** charge $-ze$
and ψ'' is the potential **at** charge $-ze$ **due to** charge $+ze$

hence

the potential ψ' **at** charge $+ze$ **due to** charge $-ze$ $= -\dfrac{ze}{4\pi\epsilon_0 r}$

the potential ψ'' **at** charge $-ze$ **due to** charge $+ze$ $= +\dfrac{ze}{4\pi\epsilon_0 r}$

where

r is the distance between the charges
each z involves only the magnitude of the charge **not** its sign
and ϵ_0 is a constant.

Charges in a Medium

Placing a charge in a material medium reduces its effect by a factor depending on the medium. This is because the electric field is now exerted through the medium, and the effect which charge $+ze$ has on a charge $-ze$ (and vice versa) is **reduced**, and the new relations include a factor ϵ_r which summarises the effect of the medium on the quantities involved.

The quantity ϵ_r is called the **relative permittivity** of the medium, and is a measure of the effect of the medium on the electrical interactions between charges (old-fashioned terminology calls it the dielectric constant).

Equations for Charges in a Medium

the electric field, E, due to the **first** charge at the position of the **second** charge $= +\dfrac{ze}{4\pi\epsilon_0\epsilon_r r^2}$

the force, F, between the two charges $= -\dfrac{z^2e^2}{4\pi\epsilon_0\epsilon_r r^2}$

the electrostatic potential energy of interaction, u, between the two charges $= -\dfrac{z^2e^2}{4\pi\epsilon_0\epsilon_r r}$

$= +ze\psi'$ or $-ze\psi''$

where the potential ψ', **at** charge $+ze$ **due** to charge $-ze$ $= -\dfrac{ze}{4\pi\epsilon_0\epsilon_r r}$

where the potential ψ'', **at** charge $-ze$ **due** to charge $+ze$ $= +\dfrac{ze}{4\pi\epsilon_0\epsilon_r r}$

where

r is the distance between the charges

z is the magnitude of the charge, **not** the sign

ϵ_0 is a constant

ϵ_r is the relative permittivity.

Index

THE POLYTECHNIC OF WALES
LIBRARY
TREFOREST